High-Pressure Shock Compression of
Condensed Matter

R.A. Graham

Solids Under High-Pressure Shock Compression

Mechanics, Physics, and Chemistry

With 98 Illustrations

Springer-Verlag
New York Berlin Heidelberg London Paris
Tokyo Hong Kong Barcelona Budapest

R.A. Graham
Sandia National Laboratories
Advanced Materials Physics Department
Albuquerque, NM 87185-5800
USA

Library of Congress Cataloging-in-Publication Data
Graham, R.A. (Robert Albert), 1931–
    Solids under high-pressure shock compression : mechanics, physics,
and chemistry / R.A. Graham.
        p.   cm.—(High-pressure shock compression of condensed
matter)
    Revised and expanded version of the author's thesis (doctoral—Tokyo
Institute of Technology).
    Includes bibliographical references and index.
    ISBN-13:978-1-4613-9280-4   e-ISBN-13:978-1-4613-9278-1
    DOI: 10.1007/978-1-4613-9278-1
    1. Materials at high pressures.   2. Shock (Mechanics).   I. Title.
II. Series.
TA417.7.C65G73   1992
620.1'1242—dc20   92-18580

Printed on acid-free paper.

© 1993 Springer-Verlag New York, Inc.
Softcover reprint of the hardcover 1st edition 1993

Production managed by Hal Henglein; manufacturing supervised by Vincent Scelta.
Typeset by Asco Trade Typesetting Ltd., Hong Kong.

9 8 7 6 5 4 3 2 1

ISBN-13:978-1-4613-9280-4

Dedication

Two hundred years ago, John Adams expressed my sentiments on priorities better than I can.

"Let me have my Farm, Family and Goose Quil, and all the Honours and Offices this world can bestow may go to those who deserve them better and desire them more. I covet them not."

This book is dedicated to Barbara, the love of my life, my constant companion and supporter, the mother of our children, "Abi" to our grandchildren. Through strength, devotion, and love of humanity, she made the lives of so many that she touched happy. Because she lived, the world is a better place.

Preface

The period from 1955 to 1985 will certainly be viewed as the golden age of shock-compression science. During that period the field grew from first recognition as a distinct scientific field with contributions from only a few highly innovative scientists to a many-faceted, broadly based discipline with major contributions to scientific knowledge and industrial technology. Scientists worldwide responded with vigor to the opportunity to study the properties of matter at unprecedented pressures and at unprecedented speeds. The scientific base grew to a world-wide science. As a result, our knowledge of geophysics, planetary physics, and astrophysics has been substantially improved with high pressure equation of state and polymorphic phase transformation data. Shock processes have become standard industrial methods in materials synthesis and processing. Shock-compression data have become the standard for the static high pressure scale. Elastic, viscoelastic, and viscoplastic deformation of solids at the highest of strain rates has been substantially defined. Large deformation elastic, piezoelectric, and dielectric properties have been studied. Chemical synthesis has been routinely carried out at unusually high pressures. Physical properties of solids have been studied at unusually high defect concentrations. Time resolution of measurement has been improved by perhaps 3 orders of magnitude. The technology has moved from remote sites to university laboratories. Characterizations of shock-compressed matter have been broadened and enriched with involvements of the fields of physics, electrical engineering, solid mechanics, metallurgy, geophysics, and materials science.

In spite of these accomplishments, shock-compression science is not documented in book form. The volumes in this series of books attempt to remedy that deficiency. The description of shock-compressed matter derived from physical and chemical observations, as presented in this book, is significantly different from that derived strictly from mechanical characteristics, which are the classical descriptions. It is hoped that the book will provide an adequate introduction for interested scientists and engineers.

This book is an edited and significantly expanded version of a dissertation presented to the Tokyo Institute of Technology in partial fulfillment of the requirements for a Doctor of Science Degree in Materials Science and Engineering. The degree was under the direction of Professor Akira B. Sawaoka. Given the press of time in an active research career, it is unlikely that the book would have been written without the requirement to write the dissertation. Much of the content of this book is developed from information in the substantial review by Lee Davison and me. That review contained over 800 references. This book adds over 100 references dated since that review.

The author has been given that superlative scientific opportunity of participating in the early development of a rapidly growing field in an unusually capable and rapidly growing research group. The solid state physics research group that nurtured the work has left its indelible mark. Certainly, this book reflects the influence of many colleagues and coauthors to whom the author is deeply indebted. They are so numerous that perhaps it is best not to attempt to list them individually. The author is deeply appreciative of the consistent and persistent support for his work from his Sandia managers and supervisors, who have performed that task with the greatest of skill and wisdom.

Contents

PART I

Shock-Compression Science

CHAPTER 1

Introduction

In this chapter: the scope of the subject; the fluidlike deformation of shock-compressed solids; modeling the shock as benign or catastrophic; the origins of shock-compression science; the pressure scale of events; the plan of the present work.

1.1 Shock Compression of Solids

Solid substances are forced into unusual and distinctive conditions when subjected to powerful releases of energy such that their inertial properties result in the propagation of high pressure mechanical waves within the solid body. The very high stress, microsecond-duration, conditions irreversibly force materials into states not fully encountered in any other excitation. It is the study of solids under this unique compression-and-release process that provides the scientific and technological interest in shock-compression science.

In extreme cases, very high pressure waves are encountered in which the time to achieve peak pressure may be less than one nanosecond. Study of solids under the influence of these high pressure shock waves can be the source of information on high pressure equations of states of solids within the framework of specific assumptions, and of mechanical, physical, and chemical properties under unusually high pressure.

In many cases, less intense pressure or stress waves are encountered in which times to achieve peak pressure may be hundreds of nanoseconds or more. The study of solids under these conditions can be the source of mechanical, physical, and chemical properties of solid materials at large strain, high pressure, and high strain rates.

Unlike gases and liquids, the temperatures induced by the rapid loading of solids can be relatively modest. Only at pressures greater than many tens of gigapascal (GPa) are the temperatures of major importance in solid density samples. In porous solid compacts, significantly higher temperatures may be encountered, but, even in this case at lower pressures, mechanical rather than thermal effects may often be dominant.

Powerful sources of energy are required to produce the transient high pressures. These include chemical energy from the detonation of high explosives, kinetic energy from the impact of high speed projectiles, kinetic energy from pulsed atomic particles, optical energy from pulsed lasers, or nuclear energy from neutrons or x-rays.

Historically, the problems studied and the approaches followed in scientific investigations are strongly constrained by the loading methods and diagnostics available to a particular investigator. Hence, the complete scientific description of shock-compressed matter often requires the interpretation of experiments from a number of independent directions that are often not consistent with each other and may contain significant ambiguities.

In shock-compression science the scientific interest is not so much in the study of waves themselves but in the use of the waves as a means to probe solid materials. As inertial responses to the loading, the waves contain detailed information describing the mechanical, physical, and chemical properties and processes in the unusual states encountered. Physical and chemical changes may be probed further with optical, electrical, or magnetic measurements, but the behaviors are intimately intertwined with the mechanical aspects of the waves.

The scientific enterprise is concerned with the identification, interpretation, and quantification of observed responses in terms of mechanical, physical, and chemical materials properties. The technological enterprise is concerned with the utilization of materials responses or distinctive shock processes.

1.2 Fluidlike Deformation of Shock-Compressed Solids

Perhaps the most distinctive and unusual response exhibited by solids under high pressure, shock-compression loading is the extraordinary ability to deform in a fluidlike manner under a wide variety of circumstances. Observation of this fluidlike deformation is a direct indication of the substantial changes brought about in solids under the conditions of mechanical loadings. Fundamental descriptions of shock-compressed solid states require knowledge of the processes involved in the transition between its initial state and the high stress state—the processes leading to fluidlike deformation. Without descriptions of these processes, theoretical treatments of shock-compression processes and descriptions of shock-compressed matter are on no firmer scientific foundation than a modern day ether theory. One could envision, for example, a mysterious substance we might call "hugonium" in solids that manifests a property leading to different degrees of fluidlike flow under high-strain-rate mechanical loading.

But, there is no need to rely on hugonium. The theory and practice of the deformation of solids under other, less intense, loadings are well developed and show that the fluidlike flow of shock deformation is the expected consequence of the motion of defects in response to applied shear stresses that exceed the shear strength of solids. In most shock loadings, the shear stresses are well in excess of that shear strength and there is certainly ample theory and experiment to qualitatively identify overall features of the defect generation and motion. Thus, we can be certain that the shock deformation at all levels above the "Hugoniot elastic limit" state results in highly defective solid materials both in the shock-compressed state and in the post-shock state. This defect state can, of course, be modified by subsequent melting or re-crystallization processes which become part of its history.

We can be qualitatively certain that the fluidlike flow of shock deformation is a consequence of motion of defects. We cannot be quantitatively certain as to the significant, detailed descriptions and consequences of these defects. Indeed, the principal unfinished business of shock-compression science is the scientific description of the defective solid in all its manifestations.

The defect question delineates solid behavior from liquid behavior. In liquid deformation, there is no fundamental need for an unusual deformation mechanism to explain the observed shock deformation. There may be super-ficial, macroscopic similarities between the shock deformation of solids and fluids, but the fundamental deformation questions differ in the two cases. Fluids may, in fact, be subjected to intense transient viscous shear stresses that can cause mechanically induced defects, but first-order behaviors do not require defects to provide a fundamental basis for interpretation of mechanical response data.

1.3 Shock-Compression Paradigms: Benign and Catastrophic

It is well accepted that fluidlike flow is observed in shock compression. In-deed, the elastic-fluid-deformation paradigm is so thoroughly entrenched in theory and practice that the question of the influence of defects has not yet come to be regarded as a fundamental mainstream issue. Given this histori-cal situation, in order to delineate and contrast the fundamental basis on which theory and practice are constructed, it is helpful to recognize that different fundamental paradigms are often encountered without statement. The "benign shock" paradigm, which has traditionally been articulated in mainstream, shock-compression science, develops theory and interprets ex-periment in terms of the perfect crystal lattice under isotropic deformation in thermodynamic equilibrium. The "catastrophic shock" paradigm [79G01, 80G01], which is rarely articulated in mainstream, shock-compression science, develops theory and experiment in terms of a defective crystal lattice under heterogeneous, anisotropic deformation not in thermodynamic equilibrium.

The benign shock paradigm is clearly an approximation, but one that has proven very effective. The catastrophic shock paradigm corresponds to the known physical, mechanical, and chemical processes, but its characteristics have proven difficult to quantify.

In large measure the paradigm within which work is carried out is strongly influenced by the objectives of the work, the background of the investiga-tor, and the particular materials model under study. From a strictly fluid mechanics, hydrodynamic, or continuum framework, defect issues are not overtly at issue. From a strictly mechanical framework, the defective solid issues are notable but not dominant. From a solid state physics framework, optical, electrical, and magnetic effects under shock compression may find the defect issues to be very significant, perhaps of first order. From a solid state chemistry framework, defect issues are of first order.

Because of the importance of characteristic solid properties and defects, an equivalence between shock-compressed liquids and shock-compressed solids cannot be assumed. This book focuses on solids as substances with characteristics distinctively different from liquids.

A description of shock-compressed solids in terms of the catastrophic shock paradigm was stated many years ago by Kormer [68K02] as:

"The shock wave is a powerful generator of defects, formed during the strong plastic deformation taking place at the wave front. These disturbances of the ideal crystal lattice, as under normal conditions, determine to a large extent the electrical, optical, and other physical characteristics of the material. The generation of imperfections brings about an acceleration of phase transformations and is the reason for the relatively high conductivity, the absorbing power, and possibly the polarization of shock-compressed dielectrics reported by a number of investigators."

The fluid mechanics origins of shock-compression science are reflected in the early literature, which builds upon fluid mechanics concepts and is more concerned with basic issues of wave propagation than solid state materials properties. Indeed, mechanical wave measurements, upon which much of shock-compression science is built, give no direct information on defects. This fluids bias has led to a situation in which there appears to be no published terse description of shock-compressed solids comparable to Kormer's for the perfect lattice. Davison and Graham described the situation as an "elastic fluid approximation." A description of shock-compressed solids in terms of the benign shock paradigm might perhaps be stated as:

"The shock-compression pulse carries a solid into a state of homogeneous, isotropic compression whose properties can be described in terms of perfect-crystal lattices in thermodynamic equilibrium. Influences of anisotropic stress on solid materials behaviors can be treated as a perturbation to the isotropic equilibrium state."

1.4 Origins of Shock-Compression Science

There are numerous early scientific works concerning the presence of shock waves and the influence of explosions, impacts, and shock waves on matter. The earliest work, however, did not lead to a delineation of the phenomenon as a distinct scientific enterprise. This distinction rests with a group of visionary scientists assembled at Los Alamos for the development of the atomic bomb during World War II. Having learned the methods and developed the technology to explosively load samples in a precise and reproducible manner, they realized that they had in their hands, for the first time, the ability to study matter in an entirely new range of pressure. After several precursor publications beginning in 1955, the existence of the new scientific field was reported to the world in the classic work by Melvin Rice, John Walsh, and

Robert McQueen of the Los Alamos Scientific Laboratory in their 1958 publication in the *Solid State Physics* book series [58R01]. Almost overnight, ordnance laboratories throughout the world were able to convert technology developed for weapons into technology for visionary studies of matter in a new regime. Swords forged for nuclear weapon development were beaten into high pressure science plowshares.

Spurred on by these scientific developments, the subsequent 5- to 7-yr period set the stage for research to be carried out for the next 30 yr. Publications from a nuclear weapons laboratory in the Soviet Union led by Al'tshuler began to appear in 1958 [58A01]. Publications began to appear from Stanford Research Institute in a group led by Duvall [55D01], and work was reported from Lawrence Livermore Laboratory [56A01], the U.S. Naval Ordnance Laboratory [55M01], and from Great Britain [48P01]. The early origins of solid state physics under shock compression were published in 1957 [57N01, 57A01] by the Sandia Corporation (later Sandia National Laboratories) group led by Neilson.

Capabilities in materials processing and synthesis technology were demonstrated by Duvall of Stanford Research Institute in 1960 [60D02], and quickly followed by the report of diamond synthesis from shock-compressed graphite by DeCarli and Jamieson [61D01] of that same group. The first shock-induced chemical synthesis was reported in Japan by Kimura in 1963 [63K03]. Recognizing the beginning of a "new scientific trend," Batsanov and co-workers [65B01] and Adadurov and co-workers in the Soviet Union [65A02] published the first work of a major sustained effort in shock-induced solid state chemistry in 1965. In the next year, the DuPont group reported major efforts in shock modification of ceramic materials [66B01].

The various publications show that essentially all the original significant questions concerning shock-compressed matter had been raised by 1966. The principal focus of most of the work in the intervening period was the use of shock compression to study high pressure "equations of state." Such work was, and is, carried out in the "benign shock" approximation. That solids have strength that can complicate the measurements and their interpretations was recognized from the beginning [55M02, 58R01]. Nevertheless, treatment of these "strength effects" has effectively been used as a patch placed on a fundamentally incomplete materials model.

The difficulties in dealing with these fundamental solid behaviors were aptly characterized as "the metallurgical mud" by Walsh and Taylor [84T01]. It has proven to be a difficult task to extricate shock-compression science from the metallurgical mud. Mixing of chemical ooze into the metallurgical mud has now further clouded our scientific knowledge of the processes.

1.5 Pressure Scale of Events

Use of terms such as "shock-compression science," or "shock-compression processes" casts such a broad net that little technical communication is accomplished. Within the overall framework of interest in materials under rapid impulsive loading, the processes encountered, their magnitudes, and their consequences are diverse. Thus, it is instructive before becoming involved in detailed treatments to briefly consider pressure scales of interest and some of the mechanical, physical, and chemical events involved.

To arrive at a perspective on magnitudes of pressure, consider two types of loadings, planar impact and planar detonation of high explosives, which are perhaps the two most common procedures. Figure 1.1 shows shock-

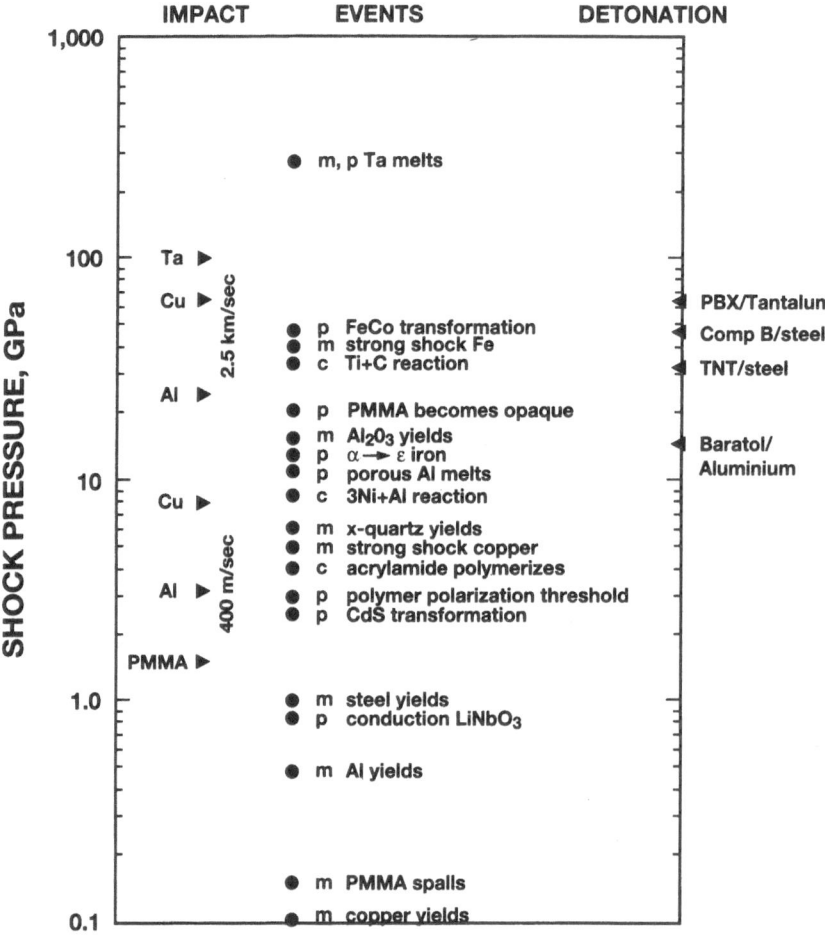

Fig. 1.1. The pressure range over which shock-compression events are of interest is very broad. Quite different and distinctive behaviors are to be expected at the various pressures. The figure shows pressures produced by impact and detonation as well as physical (p), mechanical (m), and chemical (c) events at selected pressures. The indicated impact pressures are those for impactor and target materials which are the same.

compression pressures resulting from planar impacts of solids upon each other (called symmetric impacts) at velocities of 400 m s^{-1} and 2.5 km s^{-1} and pressures produced from detonation of high explosives in intimate contact with several solids. Along with these pressures, various events are indicated. A given event, such as mechanical yielding, is shown to vary from perhaps 100 MPa to 20 GPa in different materials. Also indicated in the figure are physical and chemical events at which critical pressures are descriptive. From an examination of the various events and pressures, one should note that the pressures of interest range over 3 orders of magnitude. The conditions encountered and dominant processes are often qualitatively and always quantitatively different at the various pressures.

The emphasis of the present work is science and technology in the laboratory. The natural shock-compression laboratory of meteoritic impact should not be overlooked. In these environments unique solid state materials have been synthesized for the first time. Perhaps the most common features of our Earth, Moon, and other planets and moons are the craters produced by such high velocity impacts [67C01, 87A03].

1.6 Review Articles

There are numerous reviews of the various aspects of shock-compression science; a large number of the references were collected and summarized in Davison and Graham [79D01]. Those general reviews summarized in Table 1.1 provide an extensive source of concepts and data on materials response, and the serious student should study them carefully.

Perhaps of more interest are also numerous reviews of specialized topical areas within shock-compression science as tabulated in Table 1.2. These specialized reviews contain much detailed information on the topics under con-

Table 1.1. General reviews of shock-compression science (after Davison and Graham [79D01]).

Reference		Pages	References	Remarks
Rice et al.	[58R01]	63	41	Classic first, most widely cited
Duvall	[61D02]	37	78	General, dated
Duvall and Fowles	[63D02]	82	145	Comprehensive
McQueen	[64M01]	86	64	Author's own work
Al'tshuler	[65A01]	39	169	Soviet literature
Doran and Linde	[66D01]	61	237	Broad, uncritical
McQueen et al.	[70M01]	124	52	Author's own work, thorough
Jones	[72J01]	23	32	Elementary
Murri et al.	[74M01]	163	527	Comprehensive
Davison and Graham	[79D01]	73	>800	Broad, thorough
Asay and Kerley	[87A04]	31	128	Broad, terse

Table 1.2. Reviews of topical areas of shock-compression science (after Davison and Graham [79D01]).

Geophysics		
Ahrens et al.	[69A02] (74, 67)	
Ahrens	[72A01] (30, 124)	
Stöffler	[72S04] (63, 256)	
Grady	[77G01] (50, 91)	
Boslough	[91B01] (30, 162)	
Numerical methods		
Wilkins	[64W01] (53, 17)	
Herrman and Hicks [73H02] (34, 32)		
Johnson and Anderson [87J01]		
Zukas	[90Z01] (31, 645)	
Phase transformations		
See Chap. 2		
Optical properties		
Kormer	[68K02] (25, 178)	
Equations of state		
Knopoff [63K01, 63K02] (18, 30; 16, 35)		
Al'tshuler and Bakanova [69A01] (12, 45)		
Royce	[71R01] (16, 44)	
Royce	[71R02] (11, 11)	
Duvall	[73D01] (32, 21)	
Viscoplastic response		
Hopkins	[61H01] (14, 136)	
Wilkins	[64W01] (53, 17)	
Herrmann	[69H01] (54, 101)	
Herrmann and Nunziato [73H01] (158, 81)		
Herrmann	[76H01] (26, 54)	
Materials synthesis and processing		
Gourdin	[86G04]	
Meyers et al.	[88M01] (43, 100)	
Thadhani	[88T01] (56, 86)	
Composites		
Bedford et al.	[76B01] (54, 83)	

Experimental technique
See Chap. 3

Metallurgical effects and metalworking
Dieter	[62D01] (16, 100)
Appleton	[65A03] (6, 59)
Zukas	[66Z01] (19, 61)
Otto and Mikesell [67O01] (44, 65)	
Crossland and Williams [70C01] (21, 95)	
Leslie	[73L01] (76, 100)
Gray	[90G01] (8, 26)

Magnetic properties
Royce	[71R03] (13, 40)

Electrical conductivity
Kormer	[68K02] (25, 178)
Styris and Duvall [70S01] (22, 75)	
Keeler	[71K02] (20, 44)
Yakushev	[78Y01] (16, 65) Technique

Shock-induced electrical polarization
Mineev and Ivanov [76M01] (19, 148)

Shock viscosity
Miller and Ahrens [91A01] (30, 82)

Solid state chemistry
Adadurov et al. [73A01] (12, 55)	
Dremin and Breusov [68D01] (11, 92)	
Adadurov and Gol'danskii [81A01] (18, 53)	
Graham et al.	[86G01] (18, 45)
Graham et al.	[86G02] (26, 92)
Graham et al.	[88G01] (6, 24)
Graham et al.	[89G01] (8, 40)

Viscoelastic response
Nunziato et al. [74N01] (108, 217)

Numbers in parentheses indicate number of pages and number of references.

sideration. Some caution should be exercised on whether the information within a review is outdated.

The series of *Proceedings of the American Physical Society Topical Conferences on Shock Compresion of Condensed Matter* provides an especially rich collection of data by many authors. Table 1.3 gives a summary of those reference sources.

Table 1.3. Proceedings, American Physical Society topical conferences on shock waves in condensed matter.

Conference	Location	Editors	Publisher	Publ. date
Second, 1981	Menlo Park	Nellis, Seaman, and Graham	AIP	1982
Third, 1983	Santa Fe	Asay, Graham, and Straub	North-Holland	1984
Fourth, 1985	Spokane	Gupta	Plenum	1986
Fifth, 1987	Monterey	Schmidt and Holmes	North-Holland	1988
Sixth, 1989	Albuquerque	Schmidt, Johnson, and Davison	North-Holland	1990
Seventh, 1991	Williamsburg	Schmidt, Dick, Forbes, and Tasker	North-Holland	1992

The recent book by Young [91Y02], *Phase Diagrams of the Elements*, presents an authoritative and comprehensive account of the influence of pressure on polymorphic and solid–liquid transitions. Both theoretical and experimental work are succinctly summarized.

1.7 The Layout of this Work

To develop a terse, broad description of mechanical, physical, and chemical processes in solids, this book is divided into five parts. Part I contains one chapter with introductory material. Part II summarizes aspects of mechanical responses of shock-compressed solids and contains one chapter on materials descriptions and one on experimental procedures. Part III describes certain physical properties of shock-compressed solids with one chapter on such effects under elastic compression and one chapter on effects under elastic-plastic conditions. Part IV describes work on chemical processes in shock-compressed solids and contains three chapters. Finally, Part V summarizes and brings together a description of shock-compressed solids. The information contained in Part II is available in much better detail in other reliable sources. The information in Parts III and IV is perhaps presented best in this book.

This book will take the reader from the simplest condition of shock-compressed matter—the large elastic strain—through the complications introduced by rapid plastic deformation, to perhaps the most complex conditions—chemically reacting solids. Even in the simplest case, unexpected complexities are observed. The full complexity of shock-induced solid state chemistry is yet to be determined.

In no case is the information presented in the book comprehensive. Basic ideas are introduced and placed in perspective. Little mathematical description of processes is developed. The issues of mechanical response are afforded the least depth, as that subject has been treated in detail by numerous authors. The mainstream shock-compression area of equation of state is discussed only briefly in this book, either theoretically or experimentally. Numerous studies have been presented and much data have been tabulated in this area [80M01, 77V01]. In all cases selected reference sources are given for further study by the reader.

1.8 A Note on Sign Conventions

The information presented in this work builds upon developments from several more established fields of science. This situation can cause confusion as to the use of established sign conventions for stress, pressure, strain and compression. In this book, those treatments involving higher-order, elastic, piezoelectric and dielectric behaviors use the established sign conventions of tension chosen to be positive. In other areas, compression is taken as positive, in accordance with high pressure practice. Although offensive to a well structured sense of theory, the various sign conventions used in different sections of the book are not expected to cause confusion in any particular situation.

PART II

Mechanical Response of Shock-Compressed Solids

CHAPTER 2

Basic Concepts and Models

In this chapter: the regimes of mechanical response; nonlinear elastic compression; stress tensors; the Hugoniot elastic limit; elastic-plastic deformation; hydrodynamic flow; phase transformation; release waves; other mechanical aspects of shock propagation; first-order and second-order behaviors.

The foundation of shock-compression science is based upon observations and analyses of the mechanical responses of solid samples to shock-loading pulses. Although the resulting mechanical framework is necessary, there is no reason to believe that a sufficiently complete scientific picture can be based on mechanical considerations alone. Nevertheless, the base of our knowledge rests here, and it is essential to recognize its characteristics, and critically examine the work.

The two basic elements of a shock-compression mechanical response experiment are (i) controlled loading and (ii) careful response measurements. These aspects are illustrated in Fig. 2.1. As shown, a precisely controlled shock loading is applied over the surface of a disk, and the resulting wave profile is detected after propagation through various disk thicknesses. There are numerous methods to subject solid samples to controlled high pressure loading and to measure the wave structures produced. Physical and chemical properties can be investigated after knowledge of these observed mechanical responses is available.

This chapter presents a brief summary of principal features of wave propagation responses of solids. The review of Davison and Graham [79D01] presents a terse analysis of a significant amount of the literature in various areas of mechanical deformation to the late-1970 period. Murri et al. [74M01] describe various material deformation models, as do Herrmann and Nunziato [73H01] and Fowles [61F02]. Morris [91M01] provides a terse historical and contemporary treatment of many areas of mechanical response.

2.1 Regimes of Mechanical Response

With experimental and theoretical capabilities presently in hand, materials may be studied at peak shock stresses or pressures from perhaps 100 MPa to several TPa. This work is principally concerned with pressures from this

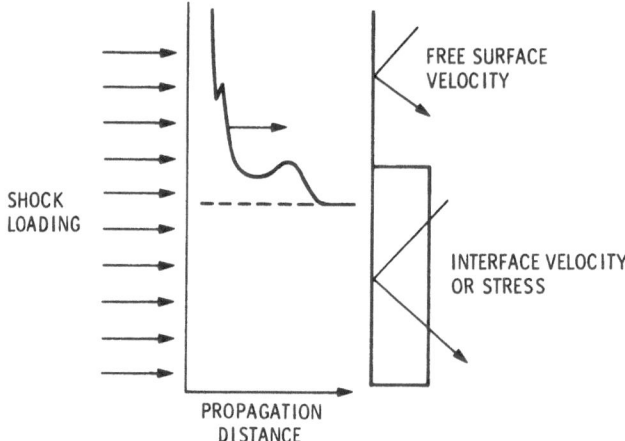

Fig. 2.1. The traditional approach to the study of mechanical responses of shock-compressed solids is to apply a rapid impulsive loading to one surface of a disk-shaped sample and measure the resulting wave propagating in the sample. As suggested in the figure, the wave shapes encountered in shock-loaded solids can be complex and may require measurements with time resolutions of a few nanoseconds.

lower stress to an upper limit of tens of GPa. Over this pressure range three typical wave profiles or wave configurations are encountered. They are: (1) a "strong shock" pressure regime, (2) an elastic stress regime, and (3) an "elastic-plastic" stress regime. In each case the response is determined by the characteristic stress-volume relation of the solid accessed by the peak applied stress.

As depicted in Fig. 2.2, each of the wave profiles is characterized by an initial powerful increase in intensity, a quiescent, but energetic, low power interlude, and a less powerful return to background conditions. In each of the regimes, Fig. 2.2 shows the corresponding stress-volume relation controlling the wave structure. Such wave profiles have been described by Duvall as having a beginning, a middle, and an end. Alternately he has thoughtfully described them as "the experimental realization of two uniform states connected by a steady transition is difficult. For this reason it is convenient to extend the meaning of "shock" to include cases where stress and density change abruptly from one value to another, even though one or both of the connected states may be changing with time" [63D02].

Strong Shock Regime

The strong shock regime is the classic archetype and is characterized by a single narrow shock front that carries the material from its initial condition into a new high pressure, elevated temperature, high kinetic energy state. Following a quiescent period at peak pressure, whose duration depends upon

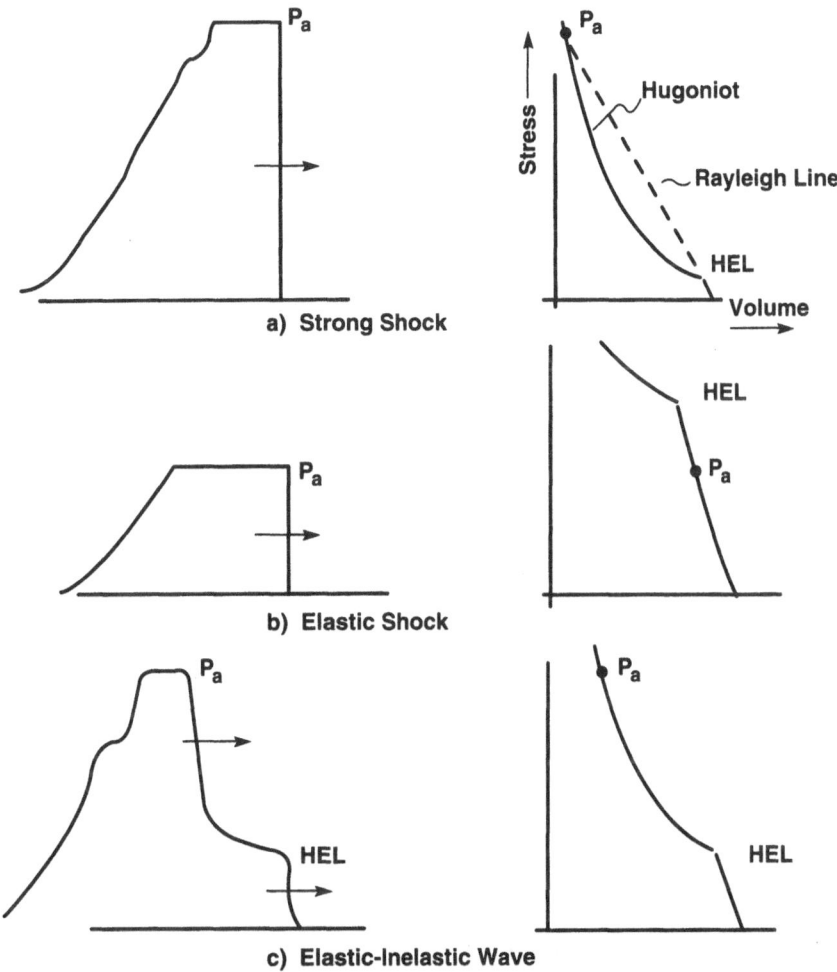

Fig. 2.2. The characteristic stress pulses produced by shock loading differ considerably depending upon the stress range of the loading. The first-order features of the stress pulses can be anticipated from critical features of the stress-volume relation. In the figure, P_a is the applied pressure and HEL is the Hugoniot elastic limit. Characteristic regimes of materials response can be categorized as elastic, elastic-plastic, or strong shock.

the loading details and the sample properties, the pressure is reduced as an isentropic unloading wave that carries the material more slowly to an atmospheric pressure, elevated temperature, and zero kinetic energy state. Even though the loading front shows no direct evidence for solid behavior, the unloading typically may show a region early in the release profile which is controlled by the strength of the material at the particular shock pressure encountered. Study of the early portion of the release wave is often described as characteristic of high pressure "sonic" behavior.

Fluidlike behavior in the high pressure shocked state may be deduced by the stress-volume values, which typically correspond reasonably closely to the same material compressed hydrostatically to the same thermal state. In order to achieve this fluidlike condition the solid must have undergone a drastic change in its structure during the loading to overcome the lattice resistance to shear deformation. This change must occur regardless of the magnitude of the shear strength of the sample. Thus, based on our general knowledge of deformation of solids, the observed fluidlike state provides direct evidence for the formation and motion of very high densities of shock-induced defects.

The unloading wave itself provides a direct measure of the strength at pressure from the shape of the release wave. Such a measurement requires time-resolved detection of the wave profile, which has not been the usual practice for most strong shock experiments.

Conservation relations are used to derive mechanical stress-volume states from observed wave profiles. Once these states have been characterized through experiment or theory they may, in turn, predict wave profiles for the material in question. For the case of a well-defined shock front propagating at constant speed U to a constant pressure P and particle velocity level u, into a medium at rest at atmospheric pressure, with initial density, ρ_0, the conservation of momentum, mass, and energy leads to the following relations:

$$P = \rho_0 U u,$$

$$1 - V/V_0 = u/U, \tag{2.1}$$

$$E = P(V_0 - V)/2,$$

where the subscript "0" indicates the conditions ahead of the shock, E is the specific internal energy, and V is the specific volume.

The pressure is to be identified as the component of stress in the direction of wave propagation if the stress tensor is anisotropic (nonhydrostatic). Through application of Eqs. (2.1) for various experiments, high pressure stress-volume states are directly determined, and, with assumptions on thermal properties and temperature, equations of state can be determined from data analysis. As shown in Fig. 2.3, determination of individual stress-volume states for shock-compressed solids results in a set of single end state points characterized by a line connecting the shock state to the unshocked state. Thus, the observed stress-volume points, "the Hugoniot," determined do not represent a stress-volume path for a continuous loading.

The line connecting the initial state to the shocked state is termed the "Rayleigh line" characterized in shock velocity as: $U = V_0[P - P_0/V_0 - V]^{1/2}$. Equations (2.1) represent propagation into undisturbed matter, but can be

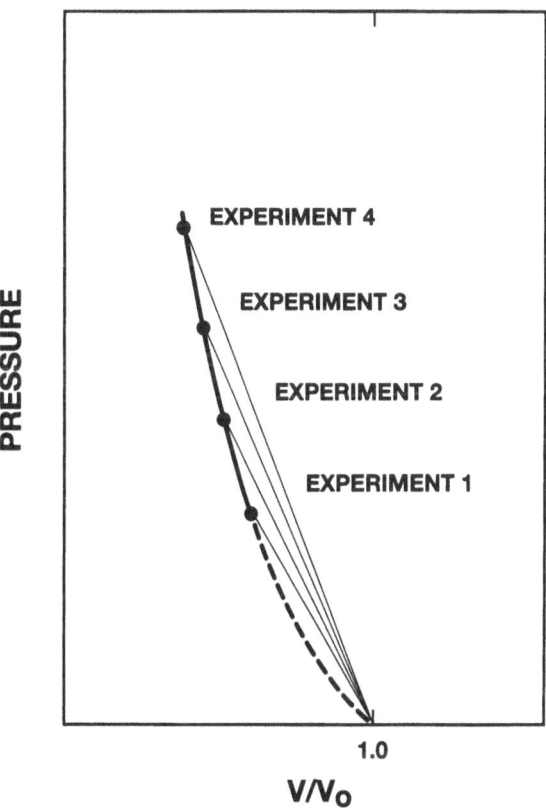

Fig. 2.3. Experimental determination of shock-stress versus volume compression from propagating shock waves is accomplished by a series of experiments carried out at different loading pressures. In the figure, the solid lines connect individual pressure-volume points with the initial condition. These solid straight lines are Rayleigh lines. The dashed line indicates an extrapolation into an uninvestigated low pressure region. Such extrapolation is typical of much of the strong shock data.

extended for multiple waves in stressed and moving matter by incrementally applying them across each wave.

Elastic Shock Regime

The elastic-shock region is characterized by a single, narrow shock front that carries the material from an initial state to a stress less than the elastic limit. After a quiescent period controlled by the loading and material properties, the unloading wave smoothly reduces the stress to atmospheric pressure over a time controlled by the speeds of release waves at the finite strains of the loading. Even though experiments in shock-compression science are typically characterized as very high pressure processes, Hugoniot elastic limits can be quite large on the scale of more conventional processes. Stresses from a few to tens of GPa are representative.

The rise times of the elastic wave may be quite narrow in elastic single crystals, but in polycrystalline solids the times can be significant due to heterogeneities in physical and chemical composition and residual stresses. In materials such as fused quartz, negative curvature of the stress-volume relation can lead to dispersive waves with slowly rising profiles.

Within the elastic range, loading applied along "nonspecific" crystallographic directions results in propagation of both longitudinal and shear waves which may be of considerable amplitude [80C01].

Within the elastic regime, the conservation relations for shock profiles can be directly applied to the loading pulse, and for most solids, positive curvature to the stress volume will lead to the increase in shock speed required to propagate a shock. The resulting stress-volume relations determined for elastic solids can be used to determine higher-order elastic constants. The division between the elastic and elastic-plastic regimes is ideally marked by the Hugoniot elastic limit of the solid.

Elastic-Plastic Shock Regime

The elastic-plastic shock region is characterized by a peak pressure less than that for which the high pressure wave would travel at a speed greater than the speed of the elastic wave. Thus, these two regimes are separated by extension of the Rayleigh line at the Hugoniot elastic limit to the higher pressure Hugoniot. This wave-profile configuration is perhaps the most characteristic of those encountered. In these cases there is typically a distinct "precursor" whose initial wave front carries the material to a cusp in the stress-volume curve, which can clearly be identified as resulting from mechanical yielding. Because plastic deformation is inherently rate dependent, the observed "Hugoniot elastic limit" as determined from the amplitudes of the elastic precursor may depend strongly on the details of the particular experiment.

Wave profiles in the elastic-plastic region are often idealized as two distinct shock fronts separated by a region of constant elastic strain. Such an idealized behavior is seldom, if ever, observed. Near the leading elastic wave, relaxations are typical and the profile in front of the inelastic wave typically shows significant changes in stress with time.

The inelastic wave shows rise times that vary quite substantially. Recognizing that the rise time is a direct indication of the balance between the viscous response of the sample and the driving force, Grady [81G01] has analyzed and compared the effective viscosity of a range of materials. These viscosities are manifestations of the dynamic deformation controlled by the shock-induced defects, heterogeneities, and their motions.

The materials responses for which the shock-conservation relations of Eqs. (2.1) are representative are limited. For a more general structured wave profile, including the unloading wave from strong shock regions, the same conservation relations expressed in terms of the longitudinal component of stress σ and Lagrangian position h are as follows:

$$\frac{\rho_0}{\rho^2}\frac{\partial \rho}{\partial t} + \frac{\partial u}{\partial h} = 0,$$

$$\rho_0 \frac{\partial u}{\partial t} + \frac{\partial \sigma}{\partial h} = 0, \qquad (2.2)$$

$$\frac{\partial E}{\partial t} - \frac{\sigma}{\rho^2}\frac{\partial p}{\partial t} = 0,$$

where t is the time.

For simple centered waves, each level of stress propagates at a discrete speed c. In this case, the wave speed corresponds to a given increment of stress or particle velocity and is a function of stress alone. Accordingly,

$$\frac{\rho_0}{\rho^2}\frac{\partial \rho}{\partial t} - \frac{1}{c}\frac{\partial u}{\partial t} = 0,$$

$$\frac{\partial \sigma}{\partial t} - \rho_0 c \frac{\partial u}{\partial t} = 0. \qquad (2.3)$$

It should be observed that, in the most general case, interpretation of the mechanical responses requires time-resolved wave-profile measurements. As shown in Eqs. (2.2) and (2.3), direct evaluation of the response requires quantitative description of derivatives of kinetic and kinematic variables.

2.2 Nonlinear Elastic Compression

The uniaxial strain deformation state results in a stress tensor in which an isotropic stress component is superposed on a shear stress component. In such a situation, and during the short duration of the loading, solids may exhibit elastic limits approaching their theoretical shear strengths. Of particular interest are the unusually large elastic stresses exhibited by the single crystals α-quartz and sapphire. With elastic strain values of a few percent, physical property studies can be carried out at strains substantially greater than possible in other loadings. As the material is elastic, wave-speed measurements provide a unique, independent method of studying higher-order elastic behavior. This section describes the status of higher-order elastic constant characterization of shock-compressed solids. *In this section tensile stress and strain are taken as positive.*

The continuum theory of deformation of elastic solids is old and well developed [65T01, 74T01], and, in its linear version, is widely applied. Nonlinear theory is of much more recent origin. Most application of nonlinear theory has been to the behavior of highly deformable materials such as rubber or to the explanation of subtle effects observed by precise ultrasonic measurements at small strain. Shock-compression experiments present an intermediate case. Materials such as vitreous silica and crystalline quartz or sapphire remain elastic to compressive strains as large as 5%–10%, exhibiting distinctly nonlinear responses over this range, and a wide variety of materials show nonlinear elastic effects under shock compression.

The relative motion of materials points in a solid body in finite strain is best represented by a deformation gradient having components

$$F_{ij} = \delta_{ij} + \partial d_i / \partial x_j, \tag{2.4}$$

where d represents the displacement of a material point from its reference position x. For large deformation, the nonlinear strain tensor

$$\eta_{ij} = 1/2(\partial d_i / \partial x_j + \partial d_j / \partial x_i + \partial d_k d_k / \partial x_j \partial x_j) \tag{2.5}$$

is representative. The linearized tensor

$$S_{ij} = 1/2(\partial d_i / \partial x_j + \partial d_j / \partial x_i) \tag{2.6}$$

is used when strains are infinitesimal.

To describe properties of solids in the nonlinear elastic strain state, a set of higher-order constitutive relations must be employed. In continuum elasticity theory, the notation typically employed differs from typical high pressure science notations. In the present section it is more appropriate to use conventional elasticity notation as far as possible. Accordingly, the following notation is employed for studies within the elastic range: t = stress, η = finite strain, with both *taken positive in tension*.

The stress relation obtained from an expansion of the internal energy function to fourth order in the finite strain η takes the following form [79D01]:

$$t_{ij} = (\rho/\rho_R)F_{ik}F_{jl}(C_{kl} + C_{klmn}\eta_{mn} + \tfrac{1}{2}C_{klmnpq}\eta_{mn}\eta_{pq}$$
$$+ \tfrac{1}{6}C_{klmnpqrs}\eta_{mn}\eta_{pq}\eta_{rs} + \cdots), \tag{2.7}$$

where t_{ij} is the stress tensor, F_{ij} is the deformation gradient tensor, C_{ij} is the Voigt elastic stiffness constant, ρ is the mass density, and ρ_R is the reference mass density. The coefficients in this equation are functions of entropy, and are subject to a variety of thermodynamic constraints and to conditions controlled by the point symmetry of materials of interest. In addition to more conventional ultrasonic measurements at small strain, some of these coefficients have been measured in shock-compression experiments that will be described in the present section. Relative to the ultrasonic studies, shock experiments usually involve a less precise measurement at a much larger strain. Since the strains encountered in shock-compression experiments cover the entire range of elastic response, no extrapolation is involved in applications and the elastic range itself is determined.

Plane waves of uniaxial strain can propagate in any direction into an undeformed isotropic body and in certain specific directions in anisotropic bodies. If the 1 axes are chosen to correspond to one of these allowable directions, the associated longitudinal stress is obtained from Eq. (2.7) in the form

$$t_1 = (\rho_R/\rho)(C_1 + C_{11}\eta_1 + \tfrac{1}{2}C_{111}\eta_1^2 + \tfrac{1}{6}C_{1111}\eta_1^3 + \cdots), \qquad (2.8)$$

where we have adopted the Voigt condensation of subscripts for symmetric tensors: $11 \to 1$, $22 \to 2$, $33 \to 3$, 23 and $32 \to 4$, 31 and $13 \to 5$, and 12 and $21 \to 6$. In dealing with the stress components and coefficients, one simply replaces the subscripts in adjacent pairs according to the above prescription, but the strains are treated according to the prescription $\eta_1 = \eta_{11}$, $\eta_2 = \eta_{22}$, $\eta_3 = \eta_{33}$, $\eta_4 = 2\eta_{23}$, $\eta_5 = 2\eta_{13}$, $\eta_6 = 2\eta_{12}$ (similar relations hold for the linearized strain tensor S_{ij}). If, as is assumed here, the material is unstressed in the reference state, the coefficient C_1 vanishes at the reference entropy. The change in entropy occasioned by passage of a shock is of the order η_1^3, which means that the entropy dependence of the coefficients is negligible to the order of the expansion given except that the coefficient C_1 contributes to the highest-order term. This contribution is presumed to be small relative to errors in the determination of this coefficient in all work to date and is neglected in subsequent discussion. If the linearized uniaxial strain is taken to be $S_{11} = \eta_{11} - \eta_{11}^2 + \eta_{11}^3 + \cdots$, the longitudinal stress component t_1 can be expressed in the form

$$t_1 = C_{11}S_1 + \tfrac{1}{2}(C_{111} + 3C_{11})S_1^2 + \tfrac{1}{6}(C_{1111} + 6C_{111} + 3C_{11})S_1^3 + \cdots, \quad (2.9)$$

where

$$S_1 = (\rho_0/\rho) - 1 = -u/U.$$

It is noteworthy that, as pointed out by Thurston [69T01], the nonlinear material responds to finite uniaxial strain (to fourth order) as though it were linear if $C_{111} = 3C_{11}$ and $C_{1111} = 15C_{11}$. A compression wave will propagate as a shock only if $C_{111} + 3C_{11} \leq 0$ (assuming the fourth-order term is negligible). This is the case with most materials and, when it prevails, the elastic coefficients can be obtained by fitting Eq. (2.9) to stress-strain states obtained from shock-compression experiments conducted over the range of elastic response or from a single experiment in which the continuum of states realized in a centered decompression waveform is recorded. Graham [72G02], using data obtained by Barker and Hollenbach [70B01], has characterized z-cut sapphire to fourth order by this method.

If $C_{111} + 3C_{11} > 0$, a centered simple wave will be produced by impact loading, and a record of this waveform suffices to determine the entire uniaxial stress-strain relation over the range of strains encountered. Vitreous silica is a material responding in this manner, and its coefficients have been determined by Barker and Hollenbach [70B01] (see also [72G02]) on the basis of a simple-wave analysis.

The simple wave produced by impacting vitreous silica has approximately the form of a linear ramp of velocity. When this ramp wave is used to load another elastic solid placed in contact with the vitreous silica, a measurement of the resulting smooth waveform introduced into the second material can be interpreted to yield its stress-strain response [79G02].

Conner has recently extended the longitudinal stress loading investigations of vitreous silica to shear loading, and shown that within the accepted elastic range the materials deformation properties are strongly influenced by shear [88C02].

Because of the peculiar high-pressure properties of vitreous silica, it is capable of producing its characteristic low-pressure ramp wave even when loaded by contact with a detonating high explosive (see Wackerle [62W01]). This fortuitous circumstance makes certain low-pressure measurements possible in laboratories otherwise equipped only for conducting the high-pressure hydrodynamic investigations discussed in the next section.

Shock-compression experiments carried out at stresses beyond the elastic range frequently produce a single stress-strain datum at the Hugoniot elastic limit. Knowledge of this limiting point is insufficient to fully determine the higher-order elastic properties of the material, as the measured value cannot be certified to be devoid of effects of a small amount of inelastic flow. Nevertheless, it is of interest to examine such data for evidence of contributions due to fourth-order elastic effects in cases where the third-order constants have been determined ultrasonically. To do this C_{111} is evaluated from Eq. (2.3). Neglecting the highest-order term, it follows that $C_{111} = 2[(t_1 - C_{11}S_1)/S_1^2] - 3C_{11}$. Substituting tabulated values for C_{11} and the measured stress and strain at the Hugoniot elastic limit (HEL) into this equation, the higher-order constant contributions, which we call C_{111}^{HEL}, are determined. A comparison of this result to other available data is shown in Table 2.1. It is notable that in almost every case the absolute value of the third-order constant inferred from the Hugoniot elastic limit measurement is greater than the absolute value of the corresponding ultrasonically determined constant. This observation indicates that the fourth-order terms contribute significantly to the stress-strain response of these materials, even at strains of only a few percent. Relatively few measurements of higher-order elastic response have been made by the method of shock compression. They are summarized and compared to ultrasonic measurements in Table 2.2. The agreement of third-order constants determined by the two methods is, in general, quite good. The third-order constants determined for vitreous silica by shock compression are probably the most accurate available for that material, with those obtained at high temperature illustrating a unique capability of the method. The shock experiments are the only source of fourth-order constants for most of the materials.

The work of the present section shows that shock-compression experiments provide an effective method for determination of higher-order elastic properties and that, by the same token, the effects of nonlinear elastic response should generally be taken into account in investigations of shock compression (see, e.g., Asay et al. [72A02]). Fourth-order contributions are readily apparent, but few coefficients have been accurately measured.

Table 2.1. Third-order elastic constants determined from Hugoniot elastic limits (after Davison and Graham [79D01]).

Material	η_x (%)	C_{xxx}(HEL) (TPa)	C_{xxx}(ultrasonic) (TPa)	C_{xxx}(HEL)/ C_{xxx}(ultrasonic)
[111] NaCl	1.8	−0.2	−0.175	1.14
[100] MgO	2.5	−6.0	−4.9	1.22
[100] MgO	0.64–0.83	−6.2	−4.9	1.27
[100] InSb*	0.58	−0.31	1.87	
[110] InSb*	0.53	−0.59	0.90	
[100] Si	4.6–5.7	−1.0	−0.83	1.20
[110] Si	2.1–2.9	−1.8	−1.5	1.20
[111] Si	2.5	−1.6	−1.33	1.20
x-cut quartz	5.6–6.8	−0.56	−0.30	1.87
x-cut quartz	5.3–8.1	−0.54	−0.30	1.80
z-cut quartz	6.7–9.7	−1.4	−0.82	1.71
z-cut quartz	4.5	−1.2	−0.82	1.46
[111] Ge	1.35, 2.2	−1.6	−1.12	1.43
[111] Ge	2.6	−1.3	−1.12	1.2
[110] Ge	3.2	−1.6	−1.31	1.22
[110] Ge	4.5	−0.81	−0.73	1.11
a-axis CdS*	3.6	−0.30	...	...
c-axis CdS*	4.3	−0.40	...	...
[100] TiO$_2$	2.3	−4.0	...	...
[001] TiO$_2$	2.1	−2.0	...	...
x-cut CaCO$_3$	1.5	−1.0	0.579	1.75
z-cut CaCO$_3$	2.2	−0.4	0.498	0.8

* These shock data are the limiting elastic compression just prior to a polymorphic phase transformation. The x direction is taken as the propagation direction.

Table 2.2. Higher-order elastic constants (after Davison and Graham [79D01]).

Material/method	C_{xx}(GPa)	C_{xxx}(GPa)	C_{xxxx}(GPa)
Vitreous silica-GE151			
shock	77.4	+550	+11 000
sonic	77.4	+603	...
Vitreous silica-Dynasil 1000			
shock	77.5	+550	+12 000
shock	81.2	+660	+14 000
shock⁻	77.6	+592	...
	($C_{44} = +306$)	($C_{112} = +505, C_{166} = +11$)	
x-cut quartz			
sonic	86.8	−210	...
shock	86.8	−300	+7500
x-cut Al$_2$O$_3$			
sonic	49.6	+3300	...
sonic	...	+3100	...
shock	...	+3250	...
shock	...	+3300	50 000

2.3 Stress Tensors

The typical shock compression experiment involves simultaneous loading of the planar surface of a large diameter disk. Considering a material element removed from a stress-free lateral surface, the applied longitudinal force produces symmetrical deformation in a radially homogeneous, isotropic body. Each material element is subjected to the same uniaxial deformation and laterally confines its neighboring element. If the material can resist shear deformation, this uniaxial strain configuration results in shear stresses, as the longitudinal and lateral stress components are not equal. This stress configuration can be decomposed into mean pressure (P_e) and shear stress (τ) components at a given compression. (Here pressures and compressions are taken to be *positive in compression*.) The shear stress component is a direct manifestation of the shear resistance, which is limited by the strength of the solid. Accordingly,

$$P_e = 1/3(\sigma_1 + \sigma_2 + \sigma_3), \qquad (2.10)$$

and for radially isotropic and homogeneous materials,

$$\sigma_2 = \sigma_3, \qquad (2.11)$$

and

$$P_e = \sigma_1 - 2/3(\sigma_1 - \sigma_2), \qquad (2.12)$$

$$\tau = 1/2(\sigma_1 - \sigma_2). \qquad (2.13)$$

Combining Eqs. (2.7) and (2.6),

$$\sigma_1 = P_e + 4/3(\tau), \qquad (2.14)$$

and

$$\sigma_2 = \sigma_3 = P_e - 2/3(\tau). \qquad (2.15)$$

Equations (2.9) and (2.10) are representative of all isotropic, homogeneous solids, regardless of the stress-strain relations of a solid. What is strongly materials specific and uncertain is the appropriate value for shear stress, particularly if materials are in an inelastic condition or anisotropic, inhomogeneous properties are involved. The limiting shear stress controlled by strength is termed τ^*.

In solids of cubic symmetry or in isotropic, homogeneous polycrystalline solids, the lateral component of stress is related to the longitudinal component of stress through appropriate elastic constants. A representation of these uniaxial strain, hydrostatic (isotropic) and shear stress states is depicted in Fig. 2.4. Such relationships are thought to apply to many solids, but exceptions are certainly possible as in the case of vitreous silica [88C02].

In crystals, the response of the crystal to a longitudinal loading may produce deformation controlled by the crystal symmetry that is not uniaxial

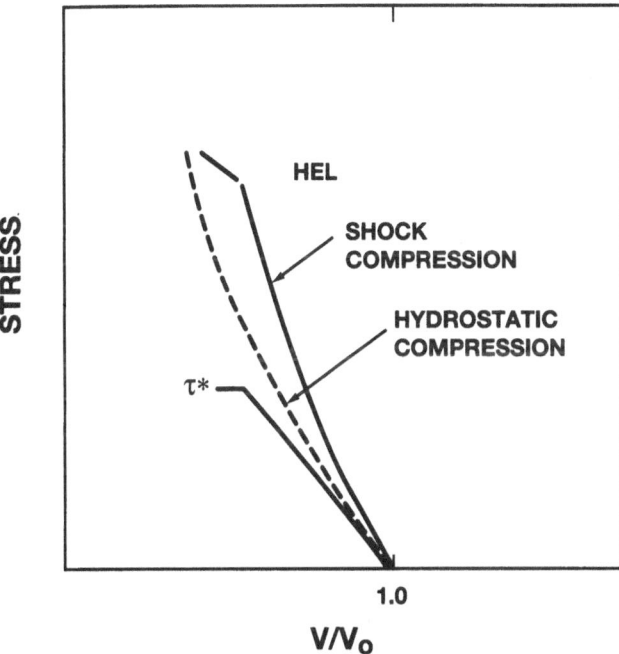

Fig. 2.4. Within the elastic range it is possible to relate uniaxial strain data obtained under shock loading to isotropic (hydrostatic) loading and shear stress. Such relationships can only be calculated if elastic constants are not changed with the finite amplitude stresses.

along the direction of the loading. Those crystallographic directions in which longitudinal loading results in uniaxial deformation along the direction of loading are called "specific directions." In other crystallographic directions, the longitudinal loading causes the propagation of quasi-longitudinal and quasi-shear waves whose amplitudes are limited by the shear strength of the crystal. These responses can be readily predicted from elastic properties of crystals [71J01].

More complete discussions of stress tensors resulting from materials strength are given by Duvall [61D02, 73D01], Jones and Graham [71J02], and Murri et al. [74M01].

2.4 The Hugoniot Elastic Limit

As loading stresses approach or exceed the shear strength of a solid, inelastic effects are to be expected, and details of the behavior have been readily observed with modern, time-resolving measurement techniques. There are many observations of these behaviors.

In a landmark conference proceedings in 1960 [61M01], Stanley Minshall of the Los Alamos Scientific Laboratory summarized detailed observations of a precursor to a phase transformation wave encountered in iron and various steel alloys. He termed the "pressure" of such waves the "Hugoniot elastic limit" [55M02] to indicate "that the (stress-volume) point was obtained by shock-wave techniques." The earlier work had termed the change in compressibility associated with the mechanical yielding phenomena as the dynamic yield point [55G01].

In that same proceedings, Duvall [61D02] tabulated the then-existing data in a table with 11 entries and provided a description of the relation between mean pressure and shear stress in uniaxial strain. Today, somewhat out-of-date tabulations [71J02, 80G03, 79D01] include entries of over 200 measurements, including metals in various metallurgical condition, brittle ceramic materials, minerals, and organics, some in both single-crystal and poly-crystalline form. Although much of the data are only a single measurement, there are a significant number of detailed studies over a wide range of materials and experimental variables.

The idealized behavior for an HEL is for a well-defined precursor shock to precede a well defined higher amplitude stress wave. This behavior is shown in Fig. 2.5. Observed elastic-plastic waveforms generally reveal the wave profiles shown in Fig. 2.6; a complex array of behaviors is noted. The time-dependent nature of the yield phenomena gives results that are sensitive to the time-resolving capabilities of the instrumentation. The behavior shown in "A" of Fig. 2.6 is thought to be typical of yielding followed by "strain hardening," that of B due to either a large strain hardening with a low amplitude

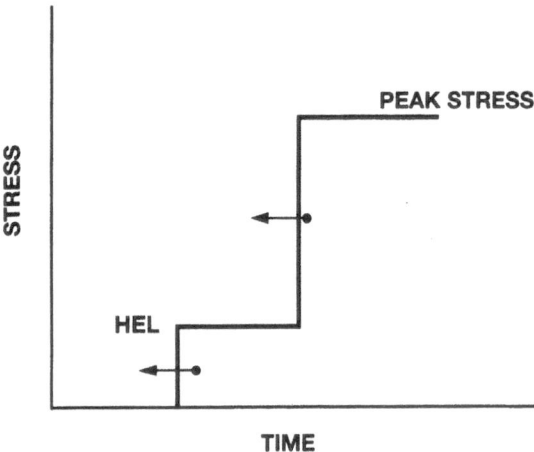

Fig. 2.5. The idealized elastic/perfectly plastic behavior results in a well defined, two-step wave form propagating in response to a loading within the elastic-plastic regime. Such behavior is seldom, if ever, observed.

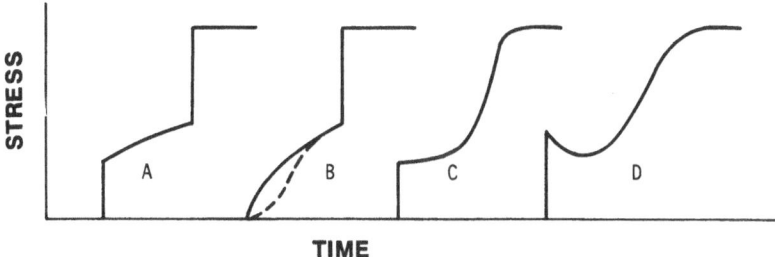

Fig. 2.6. Observed wave forms in the elastic-plastic regime are quite diverse representing nonideal elastic and viscoplastic behaviors (after Davison and Graham [79D01]).

yield stress or time-dependent yield. The wave form of C is thought to be dominated by yield followed by viscoplastic flow, while that of D is a well-defined relaxation thought to be typical of high time resolution measurements on viscoplastic flow in single crystals.

The observed behaviors are so diverse and of so many different forms that it is clear that no single model of strengths of shock-compressed solids is satisfactory. Furthermore, it should be recognized that the concept of a Hugoniot elastic limit cannot be used to quantitatively characterize a solid; rather, the term provides an umbrella expression indicating a strength value obtained under a particular set of circumstances. The one critical distinguishing feature is that the behavior is observed as a precursor to a higher amplitude wave.

A strength value associated with a Hugoniot elastic limit can be compared to quasi-static strengths or dynamic strengths observed values at various loading strain rates by the relation of the longitudinal stress component under the shock compression uniaxial strain tensor to the one-dimensional stress tensor. As shown in Sec. 2.3, the longitudinal components of a stress measured in the uniaxial strain condition of shock compression can be expressed in terms of a combination of an isotropic (hydrostatic) component of pressure and its deviatoric or shear stress component.

From the early work of Taylor [63T01] connecting dislocation behavior to observed viscoplastic shock-compression response, numerous studies have attempted to relate conventional dislocation dynamics models to experimental observations. Theory and observations consistently require unusually large numbers of mobile dislocations. Although qualitatively descriptive, progress to date on dislocation models has not proven to provide quantitative descriptions to the observations in metals.

The thorough and persistent work on "precursor decay" (the dependence of Hugoniot elastic limit on propagation distance) of Duvall's Washington State University group was successful in demonstrating that precursor attenuation was due to both stress relaxation and hydrodynamic attenuation. Typical data on crystalline LiF is shown in Fig. 2.7. Observed plastic strain

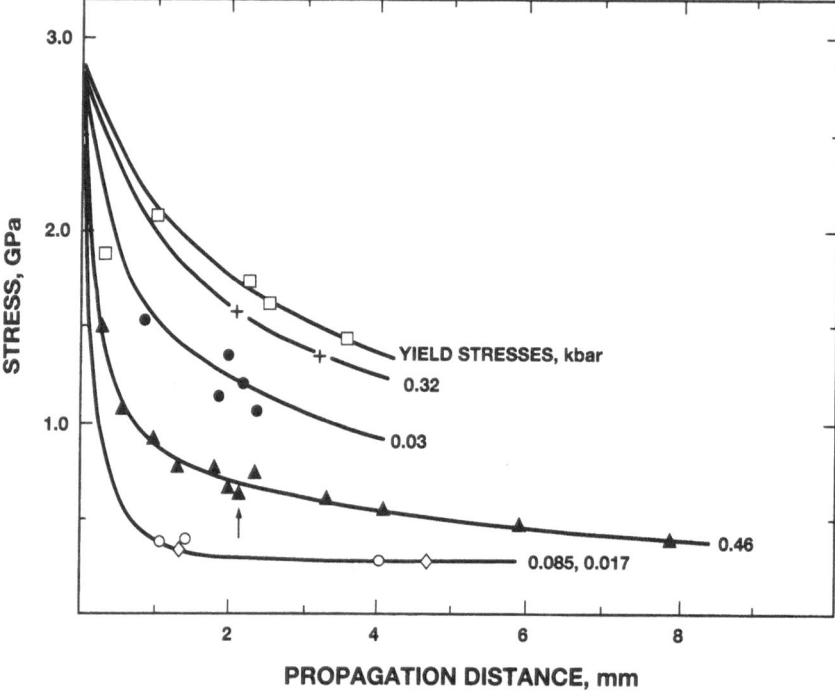

Fig. 2.7. Elastic precursor decay in which elastic waves are observed to decrease in amplitude with propagation distance is a typical behavior. The data of this figure describe the behavior of crystalline LiF samples of different yield strengths (after Asay et al. [72A02]).

controlling stress relaxation was found to be attributable to both preexisting and shock-compression-nucleated dislocations at point defects.

The idealization of a fixed shear stress at which a solid yields mechanically is often qualitatively correct, but yielding is perhaps better characterized as occurring over a range of stresses. For example, the x quartz does not exhibit a precursor until stresses exceed 6 GPa. Nevertheless, there is strong evidence that the yielding process begins to occur at stresses of 4 GPa [74G01].

There is a wealth of data and a host of models available for development of a scientific picture of yielding under high pressure shock compression. The description of the phenomena is largely phenomenological and there appears to be no accepted model or group of models that can provide a priori descriptions for the observed behaviors. For loading at stresses significantly above the HEL, wave structures reflecting typical solid behavior become quite varied, showing an array of complex viscoplastic behaviors. Table 2.3 shows Hugoniot elastic limit values for a number of characteristically different solids.

Table 2.3. Selected values of Hugoniot elastic limits.
(See Jones and Graham [71J02], Gust [80G03].)

Substance	Hugoniot elastic limit (GPa)
Sapphire, 0° orientation	20
Quartz, z-cut	10
Titanium diboride	9
Silicon carbide	8
Beryllium, c-axis	4
Beryllium, a-axis	0.4
Germanium, [100]	5
4340 steel, annealed	2
Iron, Armco	1
Aluminum alloy, 2024	0.4
Copper, [001]	0.2
Copper, annealed	0.05

2.5 Elastic-Plastic Deformation

A range of complex, elastic-plastic behaviors are observed experimentally; they are perhaps the most widely encountered and most typical of shock behaviors, but they are perhaps the least understood of the materials responses. Unfortunately, nonspecialists seldom consider realistic elastic-plastic descriptions of shock processes. This section summarizes the very large body of information available in this area. The "metallurgical mud" is most viscous in this area.

It is instructive to describe elastic-plastic responses in terms of idealized behaviors. Generally, elastic-deformation models describe the solid as either linearly or nonlinearly elastic. The plastic deformation material models describe rate-independent behaviors in terms of either ideal plasticity, strain-hardening plasticity, strain-softening plasticity, or as stress-history dependent, e.g. the Bauschinger effect [64J01, 91S01]. Rate-dependent descriptions are more physically realistic and are the basis for viscoplastic models. The degree of flexibility afforded elastic-plastic model development has typically led to descriptions of materials response that contain more adjustable parameters than can be independently verified.

In the perfectly elastic, perfectly plastic models, the high pressure compressibility can be approximated from static high pressure experiments or from high-order elastic constant measurements. Based on an estimate of strength, the stress-volume relation under uniaxial strain conditions appropriate for shock compression can be constructed. Inversely, and more typically, "strength corrections" can be applied to shock data to remove the shear strength component. The stress-volume relation is composed of the isotropic (hydrostatic) stress to which a component of shear stress appropriate to the

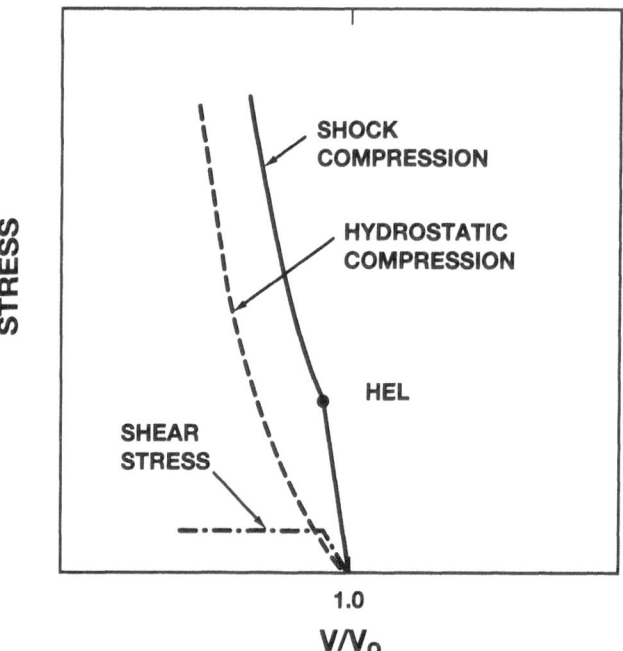

Fig. 2.8. Idealized elastic/perfectly plastic solid behavior results in a stress tensor in which there is a constant offset between the hydrostatic (isotropic) loading and shock compression. Such behavior is only an approximation which may not be appropriate in many cases.

strength of the sample is added at a given volume to equal the observed longitudinal stress. If it is presumed that the solid can be represented as having a constant strength over the range of shock pressure in question, the shock-compression behavior is simply related to the hydrostatic behavior through an offset to higher stresses by the component of strength of the solid. A stress-volume relation in such an ideal elastic-plastic approximation is shown in Fig. 2.8. The behavior shown is intuitively appealing, but can only be appealed to for heuristic purposes.

Perhaps the most dramatic exception to the perfectly elastic, perfectly plastic materials response is encountered in several brittle, refractory materials that show behaviors indicative of an isotropic compression state above their Hugoniot elastic limits. Upon yielding, these materials exhibit a loss of shear strength. Such behavior was first observed from piezoelectric response measurements of quartz by Neilson and Benedick [62N01]. The electrical response observations were later confirmed in mechanical response measurements of Wackerle [62W01] and Fowles [61F01].

This loss of shear strength was confirmed as typical of other strong solids in mechanical response studies of shock-compressed sapphire by Graham and Brooks [71G01]. In this case there was a substantial reduction, but not

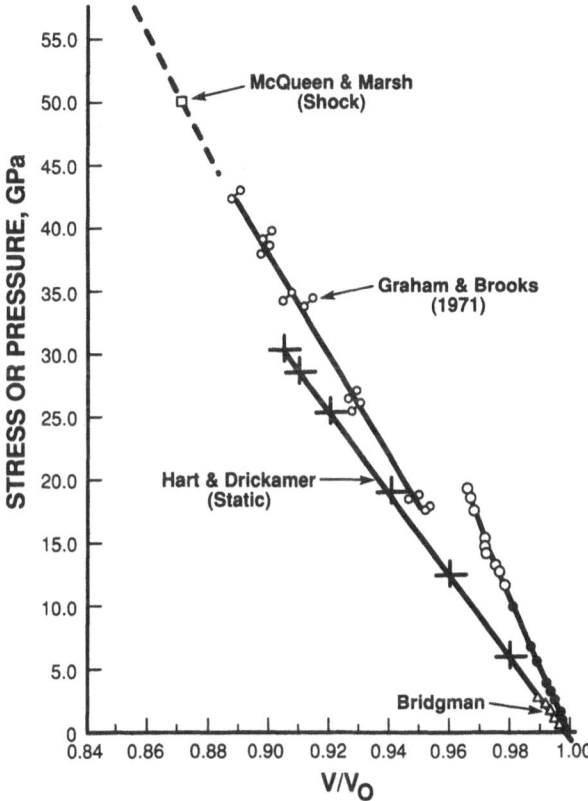

Fig. 2.9. The measured stress-volume relation of shock-loaded sapphire reveals a substantial reduction in strength, but a small finite strength is retained. The reduction in strength is indicated by the small high pressure offset between the static and shock data, and from extrapolation of high pressure shock data to atmospheric pressure conditions (Graham and Brooks [71G01]).

necessarily a full loss of strength. The observed response of sapphire is shown in Fig. 2.9. A summary of observations of loss of strength behavior in a range of materials is given in Davison and Graham [79D01], where the observations are interpreted in terms of Grady's model of localized dissipation of elastic strain energy [77G01], which leads to persistent high temperatures and local melting in low thermal conductivity solids. The idealized pressure-volume relation for such elastic-isotropic behavior is shown in Fig. 2.10.

Grady has attempted to develop concepts on viscoplastic behavior based on explicit consideration of processes within the transition zone from the elastic precursor to the high pressure loading pulse. It is observed experimentally that a large number of materials exhibit strain rates (stress pulse rise times) that vary as the fourth power of the loading pressure. This observation leads to the conclusion that the severity and high rate of the loading forces most materials to deform in similar modes. The mechanisms controlling

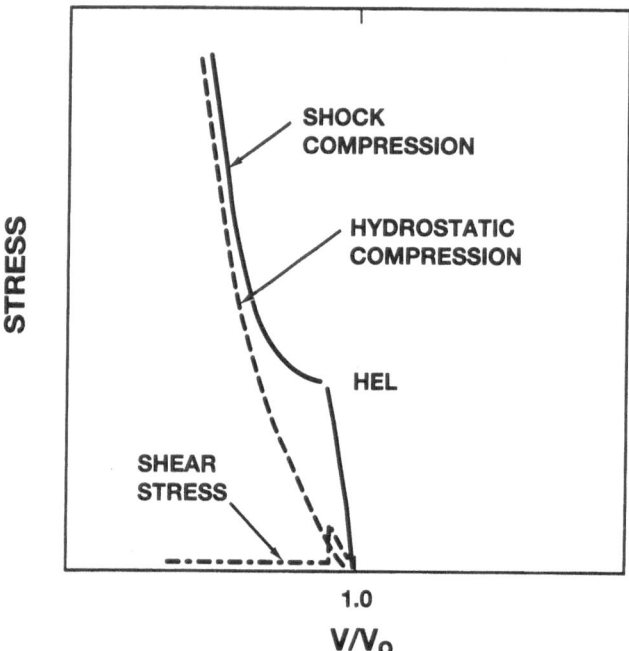

Fig. 2.10. Certain high strength solids with low thermal conductivity show a loss or reduction of shear strength when loaded above the Hugoniot elastic limit. The idealized behavior of such solids upon loading is shown here. The complex, heterogeneous nature of such yield phenomena probably results in processes that are far from thermodynamic equilibrium.

these rise times are unknown but the situation was well summarized by Swegle and Grady [85S01] as follows:

"The underlying mechanisms governing shock viscosity or the risetime of plastic shock waves, and their behavior in the shock process are not yet well understood. Viscouslike flow within the shock is thought to be associated with the microscopic processes of dislocation multiplication and motion, twinning, vacancy production, precipitate alteration, etc. Most probably it is a complex event involving the collective participation of several of these mechanisms ... and models, when coupled with numerical wave propagation codes, and with the parameters properly adjusted, have been successful in reproducing wave profile data such as those presented in this study. Such success, unfortunately, does not guarantee that either the responsible mechanisms is being modeled, or that the correct physical laws governing the mechanism have been identified...."

This statement represents an apt, terse description of the elastic-plastic shock-deformation process within the catastrophic shock paradigm.

Formalized mathematical models describing structured waves have been developed by Dunn [90D02], who also critically examined the available data. Based on this critical examination, he found that the observations, when analyzed with other viscous models, were compatible with an infinity of viscous-stress forms. Wallace has developed a large body of literature on thermomechanical behaviors and shock structures of shock-compressed solids [91W01].

Studies of the increase of shock pressure of wave profiles to some peak value provide insight into the material deformation mechanisms relating to plastic deformation and strength. Studies of the reduction of shock pressure from some peak value provide insight into solid properties at high stress states. Such studies are of technological interest to determine the wave atten-uation characteristics of solid materials. The release wave work has typi-cally been carried out to accomplish three objectives: determination of "sonic" velocity at pressure (a higher-order elastic behavior), determination of strength at pressure, and measurement of the release isentrope.

Detailed strength studies made by reloading and release of metals from high pressure provide more evidence that ideal elastic-plastic behaviors are not descriptive. An example of such a study by Lipkin and Asay [77L02] on an aluminum alloy is shown in Fig. 2.11. In this case, there is considerable

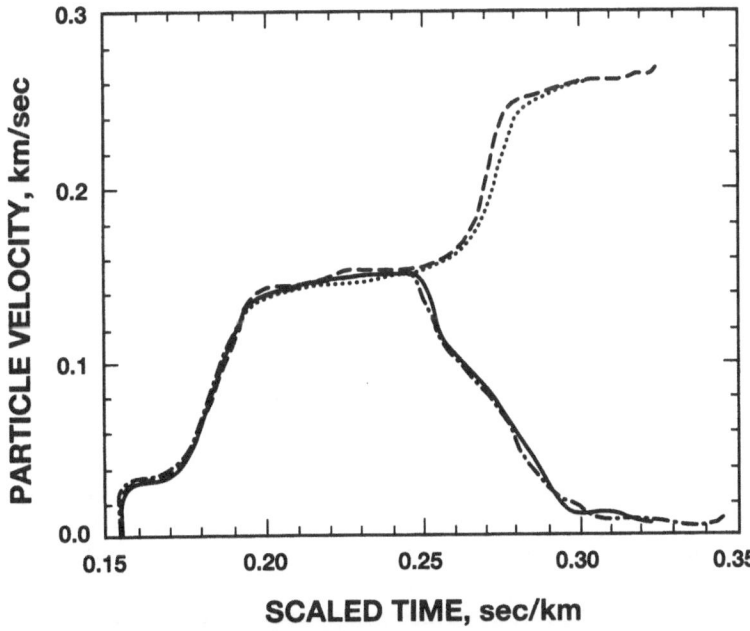

Fig. 2.11. Strength behavior of solids at pressure can be probed with reshock or release measurements. The resulting wave profiles of such measurements on a 6061-T6 aluminum alloy with VISAR instrumentation are shown. Strength behavior indi-cated on many solids reveals behavior not accurately described by simple materials models (after Lipkin and Asay [77L02]).

structure to the unloading wave indicative of viscous effects. Furthermore, the strength values obtained from release are different from the values obtained upon reloading. Release wave data are a valuable source on isentropic materials response within ideal elastic-plastic or other rate-independent assumptions.

There are no generally accepted, physically based, rate-dependent strength models, but there are numerous models such as that of Steinberg and co-workers that provide phenomenological fits to observed behaviors over a wide range in strain rates [91S01]. Follansbee and co-workers [91F01] have been somewhat successful in predicting strain-rate effects on flow stress based on a "Mechanical Threshold Stress Model" based on dislocation dynamics concepts.

2.6 Hydrodynamic Flow

With knowledge of the basis on which shock-deformation proceeds as a complex process in which fluidlike flow results from the highly defective material states, we can safely consider the deformation of solids in its simplified, idealized hydrodynamic form. In the hydrodynamic approximation, the effects of material strength are neglected or incorporated in data reduction as a "strength correction." No consideration is given to the effects of plastic deformation on the state of the material achieved. The shock state is treated as an elevated temperature, high pressure state of a perfect lattice in thermodynamic equilibrium. This approximation is justified by the resulting data on materials response that indicate that the pressure-volume states achieved do not differ greatly from those of static, hydrostatic pressure. The degree of approximation has not been well defined, and, no doubt, depends on the solid under study. Such shock Hugoniot data provide the richest source of high pressure, solid compression data available and the work certainly justifies application of the basic assumptions. The picture of solids presented is that of matter within the benign shock paradigm.

In the hydrodynamic approximation, the problem is quite well defined. Conservation of mass, momentum, and energy are used to assign pressure, volume, and energy values to a particular shock condition determined by measurement of pairs of shock velocity and particle velocity values in plane-wave high pressure loading. If the stress pulse is truly a shock, the instrumentation can be very simple and data can be produced in quantity. On this basis, hundreds of solid materials have been characterized and the data fully tabulated [77V01, 80M01]. An interesting summary of the work on copper from many authors is given in Davison and Graham [79D01]. Of particular interest here is the observation that statistical fits to the data within a limited pressure range do not extrapolate well into other pressure ranges.

The many different summaries of available data should be consulted for specific interest. The review of Davison and Graham also has a summary of different approaches to theoretical calculations of equations of state through the late 1970s. The more recent theoretical work of McMahan and co-workers is of interest [84M03]. The work on dense fluids under high pressure shock compression by Nellis and co-workers [80N01, 83N01] is a recent effort of significance.

2.7 Phase Transformations

It is well recognized that mechanical, physical, and chemical properties are strongly influenced by their crystal structures. Many solids are polymorphic; that is, they may exist in different structural forms due to the nature of the lattice bonding arrangements. Certain of these polymorphic transformations can be readily induced by the stresses and temperatures achieved in shock compression. The effect of these transformations is not only to achieve a new structure, but to substantially alter the mechanical wave profiles. There are numerous fundamental physical studies of phase stability and predictions of pressure effects. The numerous data on shock-induced phase transformations provide an excellent source of comparison of theory and experiment. As in other shock-compression science areas, the basic mechanisms driving and controlling the shock transitions are uncertain, and there are no generally agreed upon models. As we might anticipate, the basic division is between benign and catastrophic shock paradigms.

The best known of these shock-induced transformations is the 13 GPa transition in iron. Historically, it provided perhaps the first quantitative example of the potential of the shock-compression method to study high pressure materials behavior. This transition was first discovered in shock-compression studies by Minshall and co-workers [55M03] at Los Alamos. Although the features were consistent with a phase transition, it was not observed in static high pressure work of Bridgman [56B01] in pressure measurements thought to be in the same range. The shock measurements led to reexamination of the static pressure calibration scale, which was discovered to be in error.

The effect of such a transformation on a pressure-volume relation and on wave profiles is shown in Fig. 2.12. Above the transformation, its characteristics dominate the wave profile. At sufficiently high pressure, the peak pressure wave will move at higher speeds and a strong shock regime can be encountered.

When the pressures to induce shock-induced transformations are compared to those of static high pressure, the values are sufficiently close to indicate that they are the same events. In spite of this first-order agreement, differences between the values observed between static and shock compression are usually significant and reveal effects controlled by the physical and chemical nature of the imposed deformation. Improved time resolution of wave profile measurements has not led to more accurate shock values; rather,

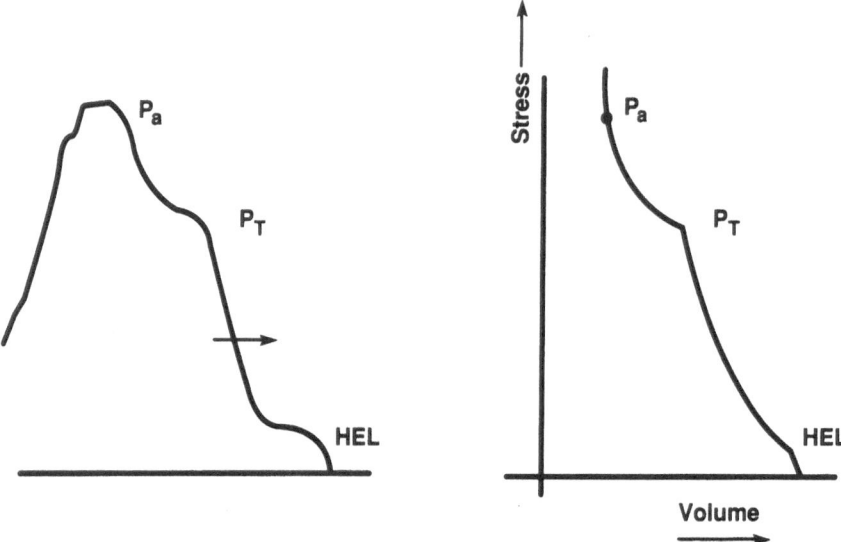

Fig. 2.12. If solids undergo a shock-induced polymorphic transformation, the volume change at the transformation causes significant changes in the wave profile produced by shock loading. In the figure, P_a is the applied pressure, P_T is the pressure of the phase transition, and HEL is the Hugoniot elastic limit.

the observations show nonideal behaviors such as are seen in elastic-plastic deformations.

Figure 2.13 shows a summary of transformation pressures in iron alloys. The present accepted value for the iron shock transition is 12.5 GPa, whereas under static pressure the accepted value is 12 GPa.

Above the critical pressure, a transformation is initiated, but, unlike isothermal equilibrium transitions, a finite pressure and volume change is typically required to complete the transition. Such a behavior is clear evidence for nonequilibrium behavior.

We can anticipate that the highly defective lattice and heterogeneities within which the transformations are nucleated and grow will play a dominant role. We expect that nucleation will occur at localized defect sites. If the nucleation site density is high (which we expect) the bulk sample will transform rapidly. Furthermore, as Dremin and Breusov have pointed out [68D01], the relative material motion of lattice defects and nucleation sites provides an environment in which material is mechanically forced to the nucleus at high velocity. Such behavior was termed a "roller model" and is depicted in Fig. 2.14. In these catastrophic shock situations, the transformation kinetics and perhaps structure must be controlled by the defective solid considerations. In this case perhaps the best published succinct statement

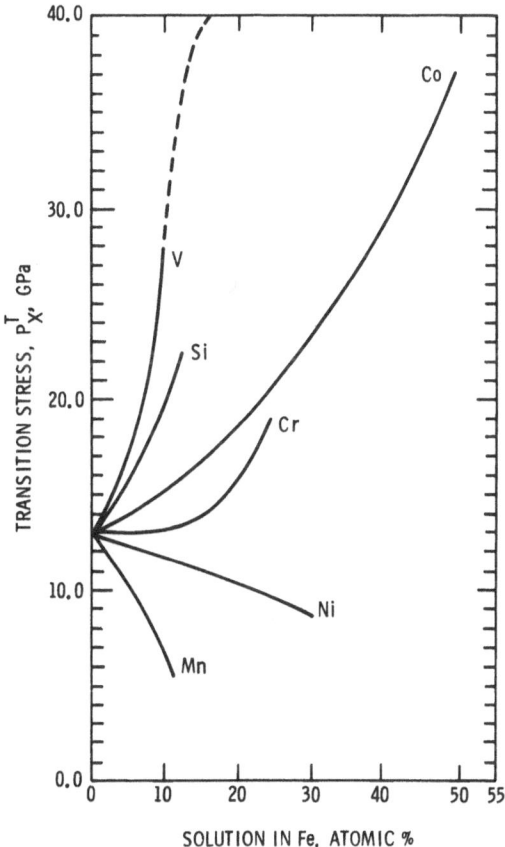

Fig. 2.13. Transition pressures for fcc iron alloys have been observed to depend strongly on the solute. The data shown represent one of the major contributions of shock-compression science as reported by Los Alamos workers (after Duvall and Graham [77D01]).

of the role of catastrophic shock effects in controlling shock-induced phase transformations is given by Al'tshuler [78A01] as follows:

"On the other hand, the formation of the high pressure phase is preceded by the passage of the first plastic wave. Its shock front is a surface on which point, linear and two-dimensional defects, which become crystallization centers at super-critical pressures, are produced in abundance. Apparently, the phase transitions in shock waves are always similar in type to martensite transitions. The rapid transition of one type of lattice into another is facilitated by nondiffusion martensite rearrangements; they are based on the cooperative motion of many atoms to small distances."

Reviews of shock-induced phase transformations are summarized in Table 2.4. The review of Duvall and Graham [77D01] emphasizes the thermo-

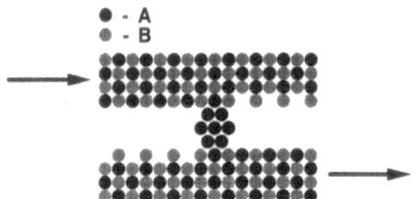

"When two layers of the substance are displaced relative to one another, the nuclei of phase A, located between them, can be regarded as kind of a roller about which oscillations are executed. - - - when the two layers of phase AB are displaced relative to one another, they transport past the nucleus, in its immediate vicinity, a multiplicity of atoms of both kinds. - - - it follows that all the atoms A passing in the immediate vicinity of the nucleus have sufficient time to combine with the latter and this in fact may be the mechanism of the growth of the nuclei of the new phase."

Fig. 2.14. Atomic level relative mass motion is an expected consequence of plastic deformation. Dremin and Breusov [68D01] have described a conceptual model of such behavior (called a "Roller Model") to explain submicrosecond structural and chemical transformations under shock compression.

Table 2.4. Reviews of shock-induced phase transformations.

Authors	Reference	Pages	References
Dremin and Breusov	[68D01]	11	92
Jones and Graham	[71J02]	12	102
Hayes	[77H01]	49	28
Duvall and Graham	[77D01]	57	405
Al'tshuler	[78A01]	10	46
Syono	[84S02]	19	95
Syono	[88S03]	8	38

dynamic characteristics of transformations and a thorough collection of data. The review of Al'tshuler [78A01] provides similar coverage and emphasizes the role of catastrophic shock effects in nucleating and controlling the kinetics of shock transformations. In contrast, the review by Syono [88S03] emphasizes the phase-transition problem in a benign shock context. Results of much of the work has had a direct effect on modeling of geophysical materials [80A01]. The work of Hayes on the KC1 transformation examines kinetics of a transformation [74H04]. The work of Tang and Gupta [88T02] incorporated the technique of imbedding an elastic crystal in a soft matrix to minimize shear effects. The book by Young [91Y02] is a valuable source of data on the influence of pressure on phase transformations.

2.8 Release Waves

Perhaps the most visible technical problems studied and the most data available on shock-compressed solids are focused on the loading portion of wave profiles. Often, the portion of the wave profile corresponding to the release of pressure to atmospheric, but elevated temperature, values is the more descriptive of solids in the high pressure state.

With nanosecond time resolution, sensitive, accurate detectors, studies of these release waves have proven to be particularly revealing. First-order descriptions of release properties were obtained with rudimentary instrumentation from the earliest studies [65A01]; it has required the most sophisticated modern instrumentation to provide the necessary detail to obtain a clear picture of the events. Characteristically different profiles are encountered in the strong-shock, elastic, and elastic-plastic regimes.

In the strong shock regime, the release wave profile provides a measure of the release isentrope. The detection of the wave speed of the leading edge or "toe" of the release is indicative of the local compressibility of the peak pressure. If the speed appears to be elastic, it is indicative of the existence of strength. Its amplitude is a measure of that strength. If the speed appears to be that of isotropic compression (bulk sound speed), and the detector is sensitive enough to differentiate from elastic speed, it is indicative of melting.

Release wave "catch-up" data taken with the optical analyzer taken for impactors of various thicknesses provide a measure of wave speed by determining the thickness of target and impactor at which the release wave catches up with the loading wave. In such an experiment the impactor serves as a controlled release-wave generator. The catch up phenomenon occurs because the release wave travels with a higher speed in the high pressure, high density state. Typical data are those of McQueen et al. [84M02] on aluminum as shown in Fig. 2.15. Interpretation of the data indicate that aluminum alloy 2024 begins to melt at 125 GPa and completes melting at 150 GPa.

Release waves in solids within the elastic regime should ideally provide a simple release fan from the high pressure state indicative of the elastic wave speed at the highest stress. Such a speed is a direct measure of the higher-order elastic constants. Shock loading and release-wave profiles obtained with the polyvinylidene difluoride (PVDF) stress-rate gauge for z-cut quartz at two stresses within the elastic range are shown in Fig. 2.16. The release-wave data show the quartz responding as a shock characteristic of a material that slightly softens up to about 5 GPa. In contrast, the shape of the release above that pressure shows considerable dispersion indicative of a slight stiffening under stress, which is the normal behavior. The observations indicate that the quartz is not fully elastic at the loading stress of about 10 GPa. Such studies also serve to demonstrate the role that sensitive wave-profile detectors can have on knowledge of subtle, but significant, processes in solid materials.

The release wave of fused quartz, thought to be elastic to stresses of

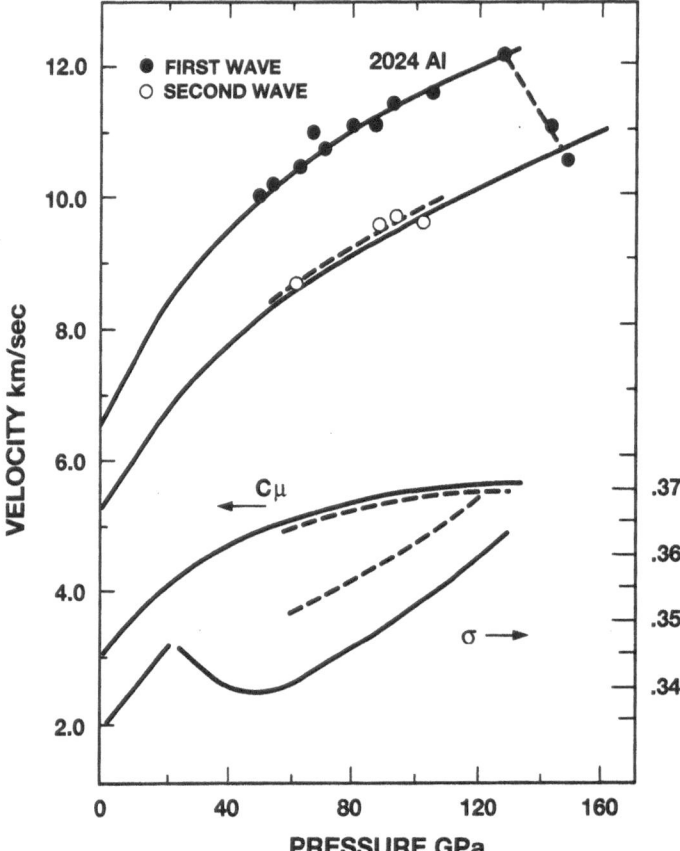

Fig. 2.15. Release wavespeeds at very high pressure can be determined by experiments in which the sample thickness is varied for fixed thickness of a high velocity impactor. Data on aluminum alloy 2024 are shown. As indicated in the figure, shear velocity (C_μ) and Poisson's ratio (σ) at pressure can be calculated from the elastic and bulk speeds if thermodynamic equilibrium is assumed (after McQueen et al. [84M02]).

10 GPa, is particularly interesting because of the anomalous slope of the compressiblity to 3 GPa. The wave profile with loading and release wave in Fig. 2.17 shows the anomalous loading and the shock on release from the high stress state.

Release waves for the elastic-plastic regime are dominated by the strength effect and the viscoplastic deformations. Here again, quantitative study of the release waves requires the best of measurement capability. The work of Asay et al. on release of aluminum as well as reloading, shown in Fig. 2.11, demonstrates the power of the technique. Early work by Curran [63D03] shows that limited time-resolution detectors can give a first-order description of the existence of elastic-plastic behavior on release.

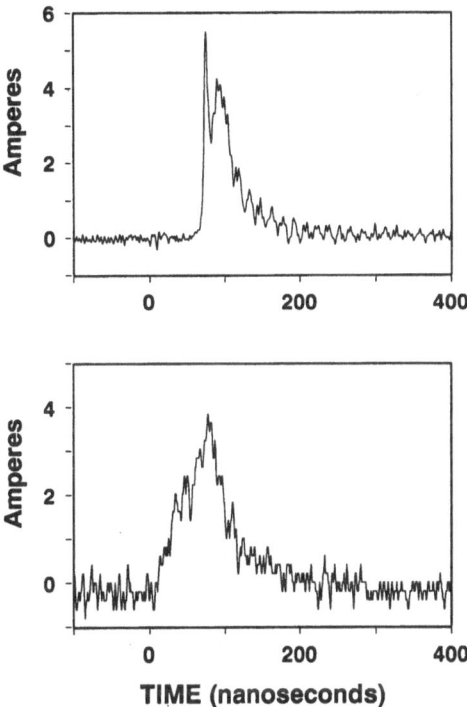

Fig. 2.16. Current-versus-time measurements with PVDF piezoelectric stress-rate gauges provide an unusually sensitive measure of detail in wave profiles. Stress-rate measurements on release waves in z-cut quartz are shown at two different peak stresses. In the figure the large positive signals are release profiles. The much smaller higher frequency signals are digitizer noise. For 4.3 GPa the release is a shock, indicating a concave downward stress-volume relation to that stress. Above that stress, the release wave signal of the lower figure is dispersed in agreement with a concave upward stress-volume relation. Time increases from left to right.

Observations of "smooth spalls" in iron provided an early, dramatic demonstration of the importance of release wave behaviors. In 1956, Dally [61E01] reported the existence of remarkably smooth fracture surfaces in explosively compressed steel. The existence of these smooth spalls was a sensitive function of the sample thickness. Analysis and experiments by Erkman [61E01] confirmed that the smooth spalls were associated with interaction of release-wave shocks and shocks from reduction of pressure at free surfaces. These release shocks are a consequence of differences in compressibility at pressures just below and just above the 13 GPa transformation.

Given the various release-wave behaviors summarized above, it is clear that release waves may often dominate wave profiles, and failure to consider their influences can lead to incorrect interpretation of observed materials responses, especially those in which samples are preserved for post-shock

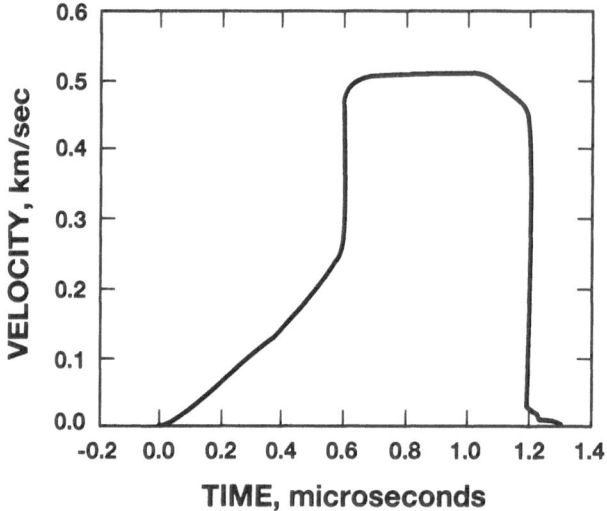

Fig. 2.17. Fused quartz is known to have an anomalous softening with stress or pressure in both static and shock loading. The time-resolved wave profile measured with a VISAR system shows the typical low pressure "ramp" followed by a shock at higher pressure. The release to zero pressure is with a shock, in agreement with the shape of the pressure-volume curve (after Setchell [88S01]).

analysis. It is common practice, for example, for work on metallurgical examination of metals preserved for post-shock study to describe "pulse duration" in simple rectangular pulse loading and unloading terms. Neither loading nor unloading is properly described in these instances.

2.9 Other Mechanical Aspects

From the mechanical response perspective, technological applications require the ability to predict wave profiles resulting from rapid impulsive loadings, while scientific studies require solution of the inverse problem of analysis of wave profiles to identify phenomena and quantify materials characteristics. As indicated earlier in this chapter, the wave profiles are shocks only in restricted stress ranges. Any phenomenon that can result in deformation can have a significant, if not controlling, effect on wave profiles. Numerous phenomena characteristic of solids can operate to substantially affect the wave profiles in subtle but physically and technologically significant ways. Neither technological nor scientific objectives can be accomplished without identification of the appropriate phenomena.

A number of such phenomena or materials characteristics are listed in Table 2.5. The noted effects include mechanical, physical, and chemical processes. The positive third-order elastic constants were described in Sec. 2.2.

Table 2.5. Phenomena causing detail in wave profiles.

Grain orientation	thermochemistry
Metallurgical condition	chemical product heterogeneity
Grain boundaries	
Precipitates	magnetostriction
Composition gradients	piezoelectricity
Elastic anisotropy	electrostriction
Melting	dynamic fracture "spall"
Viscoelasticity	porosity
Release wave dynamics	composite structure
Positive third-order constants	dynamic yielding
Structural phase transformations	
Viscoplasticity	

To complete the mechanical response description in this book, the phenomena of viscoelasticity, spall (dynamic tensile behavior), melting, and compression of porous solids are briefly considered.

Viscoelastic behaviors are strongly manifested in polymeric solids. Many Hugoniot descriptions of polymers have been tabulated, but generally ignore the consequences of viscoelasticity in spite of the fact that ultrasonic studies provide clear evidence for its influences. With the advent of time-resolved measurements such as the VISAR (see Chap. 3), studies of polymethylmethacrylate PMMA showed that the solid exhibits strong viscoelastic responses under shock loading. From measurements of Schuler [74S01], Nunziato and Walsh [74N01] developed models of viscoelastic response under shock and interpreted the data. This study serves as the archetypical work in the area.

The typical viscoelastic response, as shown in Fig. 2.18, is the propagation of a shock due to the compression, followed by a relaxation to an equilibrium state. The relaxation response is a significant part of the total response. Relaxation times are typically in the 0.1 μs regime. At pressures over about 2 GPa, PMMA shows a change in relaxation time which may be indicative of mechanical failure. Anderson has recently extended this work to other polymers and found similar strong viscoelastic behavior [92A01].

Dynamic tensile failure, called "spall," is frequently encountered in shock-loading events. Tension is created as compression waves reflect from stress-free surfaces and interact with other unloading waves or release-wave profiles. Spall has been widely studied by authors such as Curran, Ivanov, Dremin, and Davison and there is considerable data. As shown in Fig. 2.19, the wave profiles resulting from spall are characterized by an additional loading pulse after release of pressure. The late pulse is caused by wave reflection from the internal void of the tensile fracture. Analysis of such wave profiles yields appropriate spall stress values.

Alternately, samples subjected to controlled tensile loading are preserved for post-shock analysis, and sectioned so that internal cracks, their morphol-

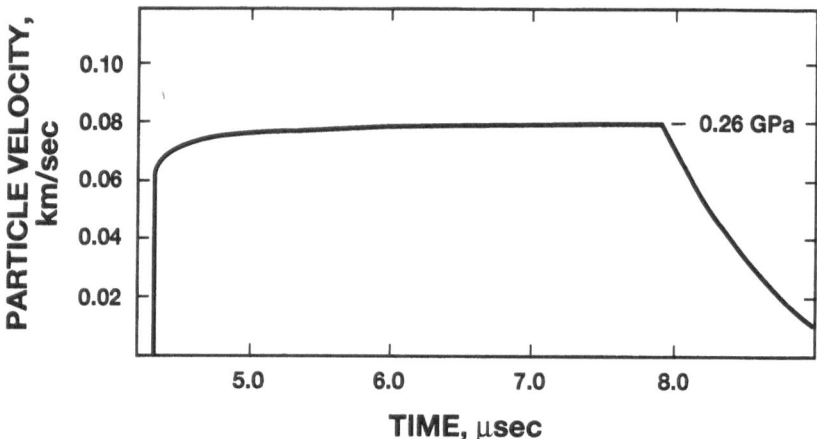

Fig. 2.18. Polymeric solids are observed to respond to shock compression in a visco-elastic behavior. The figure shows a transmitted wave profile in UVIIA PMMA as measured with an imbedded VISAR mirror. Note that the early shock is followed by a rapid relaxation to a higher velocity, and a slow relaxation to higher velocities. (after Schuler and Nunziato [74S01]).

ogy, and distribution can be examined and analyzed. Computer codes are available that incorporate various tensile failure models for prediction. (See the review by Curran [87C04].)

As in other mechanical response phenomena, simple models of a fixed spall strength appear inadequate, and there are no generally agreed upon material-response models. Beyond the usual complexities of shock-compression processes, spall introduces the additional complication of perturbation of global stresses and materials descriptions with local cracking, which can come to dominate the process. With the changes in local stresses resulting from fracture, an internal state variable model approach is perhaps the most realistic to apply [77D02], but has not been fully developed. In any event, failure to consider the consequences of spall can lead to serious discrepancies between predictions and reality.

Reviews of spall are listed in Table 2.6. Each is quite different in content and provides valuable sources of data and summaries of various material modeling efforts.

Melting, a major physical event, has small, subtle effects on shock-compression wave profiles. The relatively small volume changes and limited mixed-phase regions result in modest, localized changes in loading wave speed. Consequently, shock-induced melting and freezing remains an area with little data and virtually no information on the influence of solid properties and defects on its kinetics.

Because of the subtle effects on the loading wave profile, many of the melting studies have utilized physical property measurements such as resistivity or optical opacity. Perhaps more direct are the release-wave speed

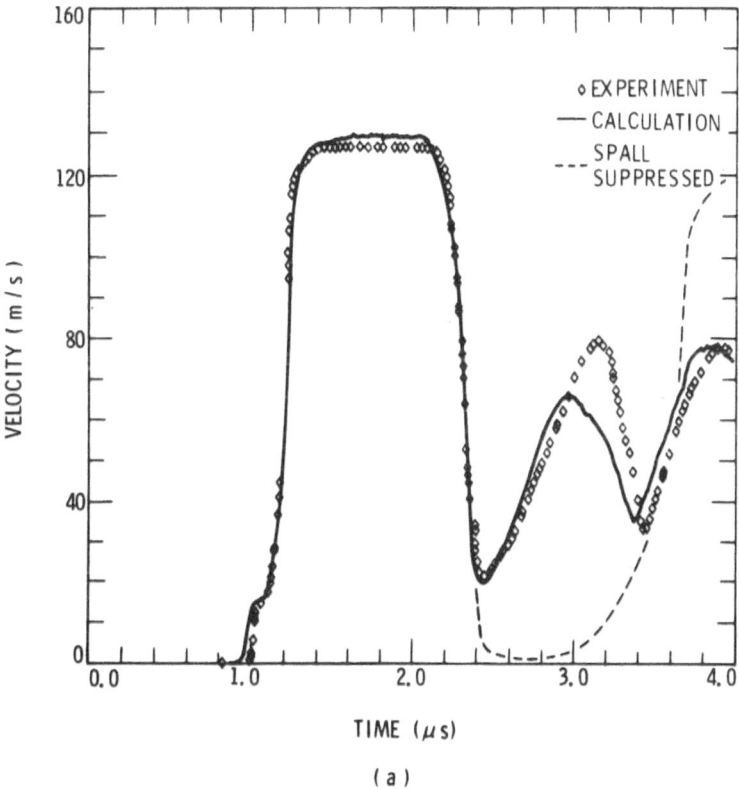

TIME (μs)

(a)

Fig. 2.19. A measured and calculated velocity versus time measurement is shown for an aluminum plate that has experienced spall failure. After the release of velocity the second and third increases in velocity represent wave reverberations within the spalled plate (after Davison and Graham [79D01]).

Table 2.6. Reviews of spall under shock loading.

Authors	Reference	Pages	References
Davison and Stevens	[71D01]	88	68
Novikov	[81N01]	10	39
Meyers and Aimone	[83M01]	91	142
Curran et al.	[87C04]	146	144
Dremin and Molodets	[90D01]	25	6

measurements. The leading edge of the unloading or "toe" travels with a speed controlled by the local slope of the stress-volume relation at the peak pressure of the experiment. In the absence of melting, this speed is the elastic longitudinal velocity at pressure. With melting, this speed is the bulk velocity at pressure. As the elastic response is relatively small in amplitude, such measurements require sensitive detectors. Two successful techniques utilize

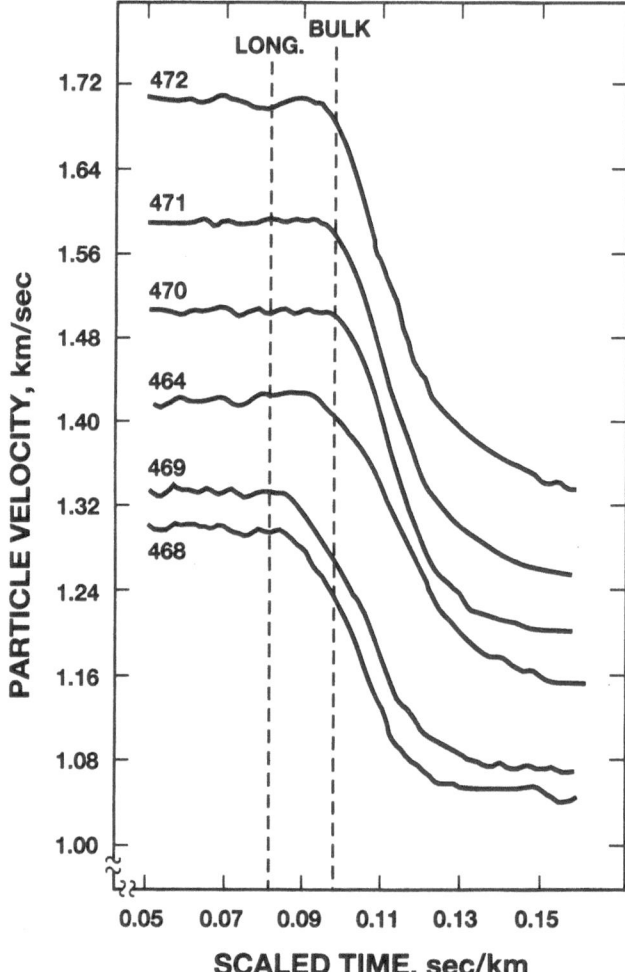

Fig. 2.20. The release wave portion of time-resolved velocity profiles in porous aluminum is shown as measured with VISAR instrumentation. At pressures near that
required to cause melt, the release changes from that of an elastic wave to that of a
bulk plastic wave, indicating the change to a melt condition (after Asay and Hayes
[75A01]).

the VISAR (see Fig. 2.20) and the optical analyzer. The data on the influence
of melting on release wave speed from Asay et al. of Fig. 2.20 obtained with
a VISAR show clearly how the speed shifts from elastic to bulk values at
shock temperatures in the vicinity of melting. Similar data on tantalum with
the optical analyzer [84B02] shown in Fig. 2.21 demonstrate the capability
of this technique to obtain data at very high pressure.

In solid density, normal melting materials, melting is typically a very high

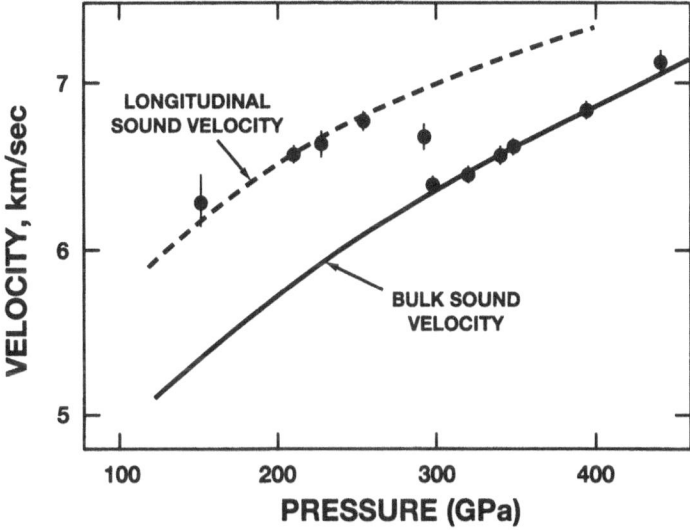

Fig. 2.21. Melting in solid density materials occurs at very high pressures. The release wave velocities measured as a function of pressure in tantalum show a shift from elastic values to bulk values at pressures approaching 300 GPa. Such a behavior is indicative of a melt (after Brown and Shaner [84B02]).

pressure event, perhaps occurring at 100 GPa or more. In porous solids, the larger volume compression accompanying the collapse of voids results in much higher temperatures and reduced melting pressures. (Note the difference in pressure scales between solid tantalum of Fig. 2.21 and porous aluminum of Fig. 2.22.)

The review of Duvall and Graham [77D01], the paper of Brown and Shaner [84B02], and the book of Young [91Y02] provide thermodynamic descriptions of the melt process.

Porous samples would appear on the surface to provide such a complex and uncontrolled local environment for deformation of solids that they would be of little interest in scientific investigations. Indeed, the principal interest in their responses is technologically driven; they are very effective attenuators of wave profiles and much of materials synthesis and processing is carried out on powders. Duvall [86D01] has summarized the difficulty of work with porous powder samples as follows:

"In summing up, I wish to emphasize the strength of my feelings that experiments of the kind which are of greatest interest to this gathering are very complicated and are exceedingly difficult to interpret in the detail required to generate new physics or to apply the known physics of shock waves. I do not imply by this that the situation is hopeless, only that understanding at a fundamental level will require intensive effort, great care, and deep thought about the processes involved. It will also benefit from collaboration with more conventional shock-wave scientists doing simpler experiments and with scientists from other disciplines like chemistry."

Nevertheless, as response data have accumulated and the nature of the porous deformation problems has crystallized, it has become apparent that the study of such solids has forced overt attention to issues such as lack of thermodynamic equilibrium, heterogeneous deformation, anisotrophic deformation, and inhomogeneous composition—all processes that are present in micromechanical effects in solid density samples but are submerged due to continuum approaches to mechanical deformation models.

As in other shock-deformation problems, naive models of shock compression of porous solids provide representative first-order descriptions of the processes. In the strong shock regime (pressures much higher than that required to compress the sample to solid density) the outstanding feature of the materials is the large contribution of heating. The location of the high pressure-volume states is controlled to first order by the initial porosity of the sample. As shown in Fig. 2.22, at sufficiently high pressure, the relationships actually show an expansion from their initial states due to the unusually large shock heating. There are numerous investigations of such high pressure behaviors and they have been effectively used to determine Grüneisen coefficients (See, e.g., [89T02]). Oh and Persson have recently proposed a new thermodynamic description of shock compression of porous solids [90O01].

In lower pressure environments, the wave profiles are dominated by the consequences of deformation of the samples to fill the voids. This irreversible "crush-up" process strongly controls the wave speeds, which have anomalously low values at low initial sample densities. Modeling of this problem is

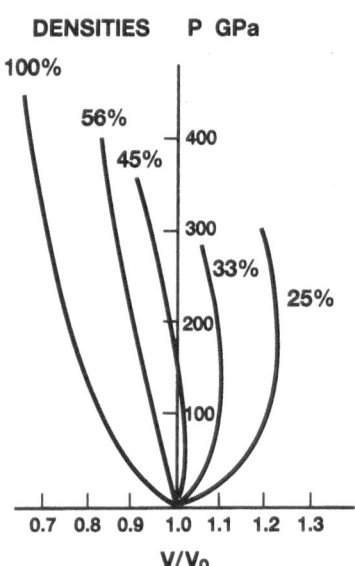

DENSITIES   P GPa

100%

56%

45%   400

300

33%

25%

200

100

0.7 0.8 0.9 1.0 1.1 1.2 1.3

V/V_0

Fig. 2.22. At very high pressures the observed pressure-volume relations of porous samples show expansions due to the very high temperatures produced in collapsing the voids. The figure shows representative behavior at porous sample densities of 25% to 100% of solid density.

most universally carried out with Herrmann's P-α model [69H02], which characterizes the process with a pressure dependent void content and leads to phenomenological description controlled by the crush pressure to achieve solid density. In this model, α is a measure of the degree of compaction. Other similar models have been used by Carroll and Holt [72C05]. Meyers has incorporated material strengths into such a modeling process with some success [89M01]. A phenomenological "snow plow" model is sometimes useful. It incorporates a crush to solid density at any elevated pressure. Most solid state chemistry problems as described in Part III of this book are carried out in porous samples. Murri and co-workers [74M01] provide a review of porous material deformation models and available data at low pressure.

2.10 First- and Second-Order Behaviors

There is a wealth of mechanical response data in all characteristic wave profile regimes: elastic, strong shock, and elastic-plastic. Less detailed data, but still considerable, are available for solids in stress regimes influenced by structural phase transformations. There are numerous data available to describe spall, viscoelastic response, phase transformations, melting, and shock compression of porous solids. Furthermore, appropriate material response models have been developed to describe the observed characteristic responses. The models have been incorporated into complex and sophisticated computer codes that can provide realistic numerical simulations of one-, two-, and three-dimensional rapid impulsive loading problems. (See the reviews by Johnson and Anderson [87J01] and by Zukas [90Z01].) These computer codes often incorporate realistic descriptions of chemical reactions appropriate for detonation of high explosives. Given the advanced state of the technology and the large supply of data, it is relevant to inquire as to the status of knowledge in the area. Is this a stagnant field with fully developed science or is there a need for continued research?

It has been a persistent characteristic of shock-compression science that the first-order picture of the processes yields readily to solution whereas second-order descriptions fail to confirm material models. For example, the high-pressure, pressure-volume relations and equation-of-state data yield pressure values close to that expected at a given volume compression. Mechanical yielding behavior is observed to follow behaviors that can be modeled on concepts developed to describe solids under less severe loadings. Phase transformations are observed to occur at pressures reasonably close to those obtained in static compression.

In spite of these representative first-order descriptions, experiments, theory, and material models do not typically agree to second order. Compressibility (derivatives of pressure with volume) shows complex behaviors that do not generally agree with data obtained from other loadings. Mechanical yielding and strength behavior at pressure show complexities that are not understood and not incorporated in existing models. Careful experiments reveal significant differences between wave profile measurements and model predictions. Phase transformation pressures differ significantly from static values, and mixed-phase regions show responses that are not accurately modeled by a process in thermodynamic equilibrium. (See, e.g., [71T01].)

Given the restrictive conditions under which theory and experiment can be accurately tested, there is reason for serious concern that the status of knowledge is purely phenomenological. There is reason for concern that observed differences between predictions and observations are due to unidentified processes and phenomena that could become significant in materials-response conditions outside those encountered in model development. There is reason for concern that the phenomenological studies rest too heavily upon parameter fittings such that observations determine parameters that, when incorporated into computer codes, yield the observations upon which they were based. To develop confidence in the fundamental descriptions of shock-compression processes, mechanical, physical, and chemical studies must be considered and integrated into an overall picture. The effort requires careful and critical examination of details of experiments and experimental apparatus. A very strong experimental and theoretical foundation has been laid for future work. Integration of the mechanical responses work with sophisticated materials science technology appears necessary to advance knowledge in the area.

CHAPTER 3

Experimental Methods

In this chapter: shock-loading methods; measurement of wave profiles; physical categories of detectors; shock-compression gauges; advances in measurement technology.

Our immediate and instinctive reaction to an impact or explosion leaves a mental image of utter chaos and destruction. There may be a fascination with the power of such events, but our limited time resolution and limited pressure-sensing abilities cannot provide direct information on the underlying orderly mechanical, physical, and chemical processes. As with other phenomena not subject to direct examination by our human senses, the scientific descriptions of shock and explosion phenomena rest upon a collection of images of the processes which are derived from a range of experiences. The three principal sources of these images in shock science—experiment, theory, and numerical simulation—are indicated in the cartoon of Fig. 3.1.

The mental images, no matter how well grounded scientifically, are individually and collectively biased, as they have been developed after considerable filtering. The filters result from the scientific training of individuals, available supporting information from other processes, existing theoretical methods, limitations of numerical simulation, and characteristics of experimental methods.

The shock-compression events are so extreme in intensity and duration, and remote from direct evaluation and from other environments, that experiment plays a crucial role in verifying and grounding the various theoretical descriptions. Indeed, the material models developed and advances in realistic numerical simulation are a direct result of advances in experimental methods. Furthermore, the experimental capabilities available to a particular scientist strongly control the problems pursued and the resulting descriptions of shock-compressed matter. Given the decisive role that experimental methods play, it is essential that careful consideration be given to their characteristics.

A number of thorough reviews on measurement techniques are listed in Table 3.1; each has a somewhat different thrust. The early review emphasizing wave profile measurement methods of Graham and Asay [78G01] has

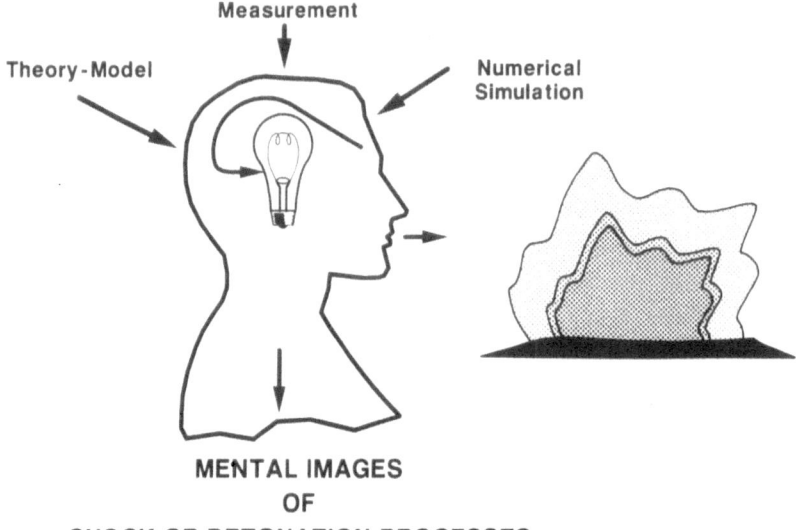

MENTAL IMAGES
OF
SHOCK OR DETONATION PROCESSES

Fig. 3.1. Mental images of shock-compression processes vary considerably depending upon the background and experience of the investigator. The scientific images are created from inputs from theory, numerical simulation, and experiment. The critical nature of the experiment in establishing "reality" requires unusually careful study of critical aspects of experimental apparatus.

Table 3.1. Reviews of wave profile measurement techniques.

Authors	Reference	Pages	References
Deal	[62D03]	2	27
Doran	[63D01]	27	57
Keeler	[71K01]	30	45
Fowles	[73F01]	74	60
Grady	[77G01]	50	91
Graham and Asay	[78G01]	36	218
Cagnoux et al.	[87C02]	74	90
Ahrens	[87A01]	50	76
Chhabildas and Graham	[87C01]	18	73

been recently supplemented by Chhabildas and Graham [87C01]. The review by Cagnoux and co-workers [87C02] emphasizes "Lagrangian Analysis" (multiple, *in situ* measurements of unsteady waves), whereas the review of Ahrens [87A01] emphasizes theory and technique for high-pressure studies of geophysical materials. Morris presents a terse summary of several important measurement techniques [91M01].

3.1 Shock-Loading Methods

The most widely used methods of applying controlled shock loading are based on plane-wave high explosive generators or precisely controlled projectile impacts. Other less useful but interesting methods include intense pulsed lasers, particle beams, and x-rays or neutrons from the detonation of nuclear devices. The loading introduced by the powerful transformation of energy into mechanical waves may be further shaped or modified by the use of wave-profile-shaping configurations placed between the loading source and the sample.

The general requirements for the loading are that the load be applied in a shorter time than that required for sample responses to be accurately measured, and also that the load be applied over the sample face in a time shorter than the same measure. With such arrangements, the sample is "shock loaded." Whether "shock waves" are produced in the sample is not under the control of the investigator; the mechanical waves in the sample are directly controlled by inertial properties with characteristic responses as described earlier in Chap. 2.

Given limits to the time resolution with which wave profiles can be detected and the existence of rate-dependent phenomena, finite sample thicknesses are required. To maintain a state of uniaxial strain, measurements must be completed before unloading waves arrive from lateral surfaces. Accordingly, larger loading diameters permit the use of thicker samples, and smaller loading diameters require the use of measurement devices with short time resolution.

High Explosive Systems

The development of high-explosive, plane-wave generators provided the initial capability used for studies of high pressure materials response. Workers at Los Alamos developed plane wave generators with diameters of 56, 102, 152, 206, and 305 mm. With the use of thick disks of high explosives such as baratol, TNT, Composition B, and PBX in direct contact with metal plates as shown in Fig. 3.2, a range of pressures can be reliably and reproducibly produced in samples (15–45 GPa in aluminum). There are considerable data available on these explosives which can be relied upon to achieve nominal driving pressures. "Cross curves" of explosives in contact with standard solid materials are shown in Fig. 3.3. Use of these high explosive systems requires firing sites located in remote areas, as the amount of explosives is substantial.

To achieve pressures intermediate to those achieved by direct contact with a given metal plate, use is often made of alternate layers of various shock impedance materials. Table 3.2 gives a summary of experimental arrangements that have been used in materials studies to achieve pressures from 3 to 80 GPa.

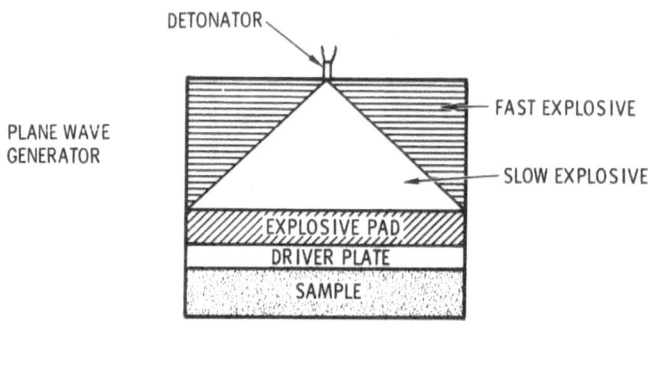

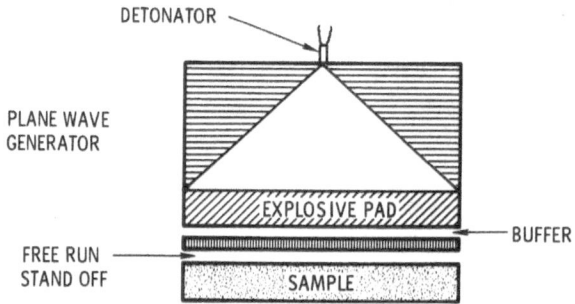

Fig. 3.2. Controlled, high pressure shock loading can be routinely carried out with large diameter plane wave explosive lenses which initiate Detonation in cylinders of high explosives with known, reproducible behavior. Detonation waves from the explosive are transmitted into metal plates which can serve as standards and on which samples to be studied are placed.

To achieve higher pressures, these same high explosive systems can be used to accelerate metal plates to high velocity [60M01]. Typical plate impact conditions are described in the Los Alamos publications cited above and given in the table.

For shock-synthesis and processing experiments, less precise systems are typically employed. These systems use commercial explosives that may be used to accelerate plates or to compress samples in the form of a tube. These systems are suitable for establishing nominal shock conditions for materials processing experiments, but are generally not suitable for careful characterization of materials response [87G02, 88M01].

Guns: Precise Impact Machines

Perhaps the most substantial change to occur in shock-compression science technology since the 1960s is the widespread use of smooth bore guns to propel projectiles to preselected velocities and impact them upon samples

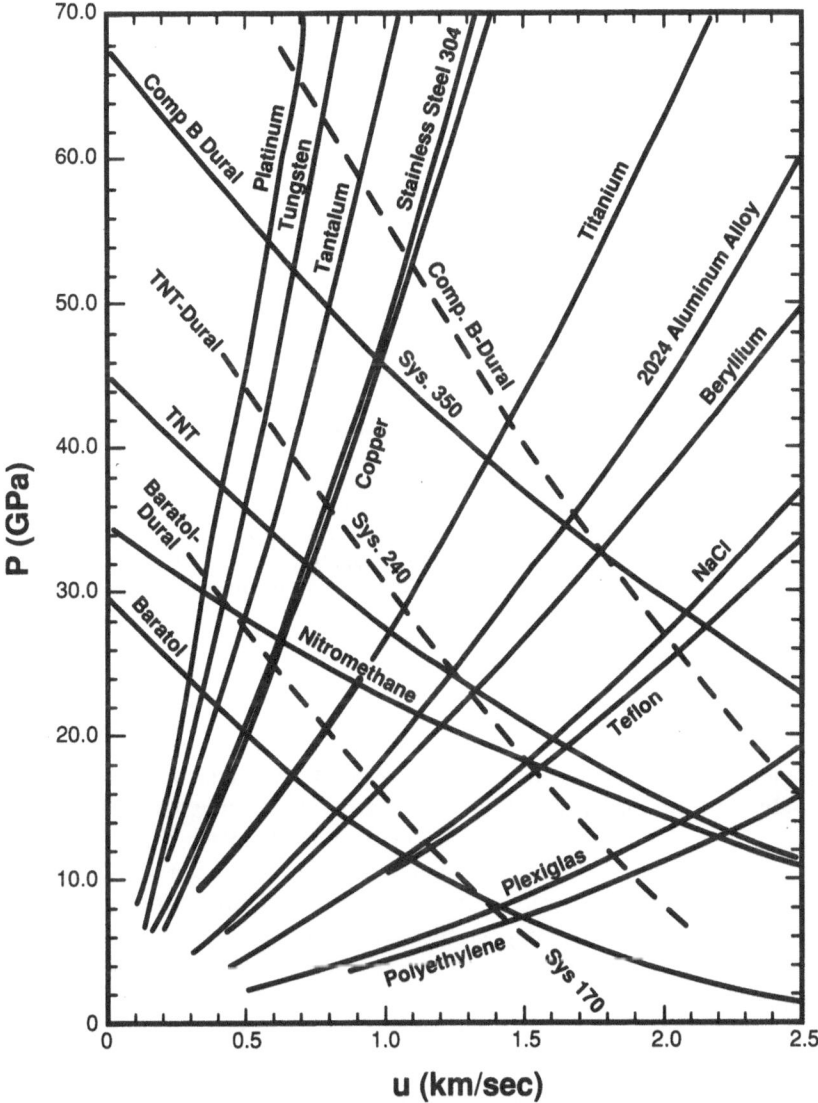

Fig. 3.3. Stress-particle velocity characterizations of many materials have been documented. The explosive "cross curves" superposed on the materials responses provide approximate loading stress levels to be determined from the intersection of the explosive and material curves. For example, the detonation of TNT produces a pressure of 25 GPa in 2024 aluminum alloy.

Table 3.2. High explosive loading systems characteristics (after Wackerle [62W01]*, Gust et al. [73G01]**).

Lens diam (mm)	HE thickness (mm)	HE type --	Driver plates/spec. plate (thickness, mm)	Pressure[†] (GPa)
203*	50.4	Baratol	PMMA(20.3); brass(25.4)/PMMA(17.3)	3.0
203*	50.4	Baratol	brass(25.4)/PMMA(17.3)	3.6
305**	152	TNT	brass(25)/PMMA(13)/Al alloy(13)	10.0
203*	102	Baratol	brass(25.4)/dural(12.7)	11.2
305**	152	PBX9205	brass(25); PMMA(13)/Al alloy(13)	14.0
203*	102	TNT	brass(12.7)/Dural(12.7)	16.6
305**	152	TNT	--/Al alloy(13)	21.8
203*	102	TNT	--/dural(12.7)	22.7
305**	152	PBX9404	brass(13)/Al alloy(13)	27.5
203*	102	Composition B	--/dural(12.7)	32.9
305**	152	PBX9404	--/Al alloy(13)	36.9
305**	152	TNT	air(3); monel(3); air(3)/Al alloy(5)	53.3
203*	102	Cyclotol	stainless(3.18); air(25)/dural(6.3)	62
305*	152	Cyclotol	stainless(4.74); air(54)/dural(32)	66
305**	152	PBX9205	air(3); monel(3); air(25)/Al alloy(5)	68
305**	152	PBX9404	air(3); monel(3); air(25)/Al alloy(5)	79
305*	152	Cyclotol	monel(2.4); air(51)/dural(6.4)	85

Other systems are documented in the cited references.
[†] Pressure at output surface of specimen plate.

with precise alignment between impacting surfaces. From modest beginnings, the impact technique has become the standard loading technique and is routinely used for the most precise experiments worldwide. Critical considerations for such impact machines include the bore diameter, the energy source used to accelerate the projectile, the velocity range that can be achieved, the accuracy of the impact velocity measurement, and, most critically, the "tilt" or misalignment between impacting surfaces. The acceleration distance of the projectile can also influence the design of impactors placed on the projectile face.

The precise impact experiment offers both versatility and the highest of precision. In typical experiments, the projectile is faced with an impactor that is either the same material as the sample to be impacted or a standard material whose properties are known and reproducible. In a "symmetric impact" configuration the impactor and sample are the same; upon impact, precisely one-half of the impact velocity is imparted to the sample. As the impact velocity can be readily measured to an accuracy of 0.1%, this condition provides the most accurately known and best characterized condition achieved in shock-compression science.

For more general impact configurations, the condition of equal particle velocity and equal pressure at impact can be used to determine the impact conditions. Typical stress versus particle velocity relations for low pressure

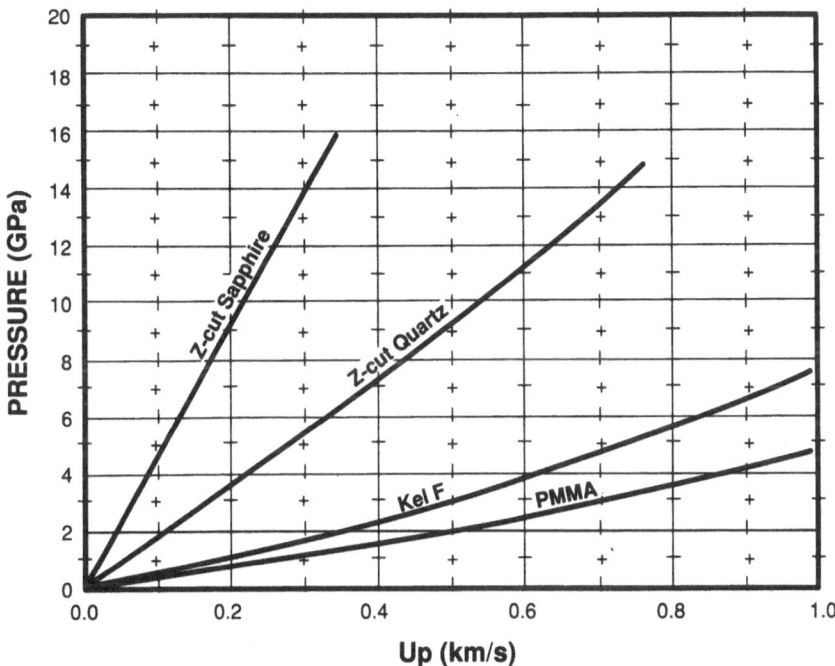

Fig. 3.4. Representative stress-particle velocity relations such as those shown in the relatively low shock-pressure region are used to determine impact stresses with good precision.

standard materials are shown in Fig. 3.4. For higher pressures, the stress-particle velocity relations of Fig. 3.3 can be employed.

Gun propellant and compressed gas are the most widely used energy sources. In single-stage, compressed-gas systems velocities can be achieved from perhaps 30 m s^{-1} to 1.5 km s^{-1}. These compressed-gas gun systems are the most widely used as they are quite safe and can be incorporated into typical university and industrial laboratories. A typical compressed-gas gun experimental configuration is shown in Fig. 3.5.

For higher velocities and large diameter bores, the higher power available from propellants is required to accelerate projectiles.

Impact velocities from about 2 to 6 km s^{-1}, which can produce shock pressures in the 110 GPa range, are routinely achieved with two-stage, light-gas guns. These systems are typically limited to diameters of from 12 to 25 mm. To achieve these velocities, propellant is used to accelerate a large piston which serves to compress a reservoir of helium or hydrogen gas to high pressure. The gas then accelerates the projectile over an acceleration distance of perhaps 25 m.

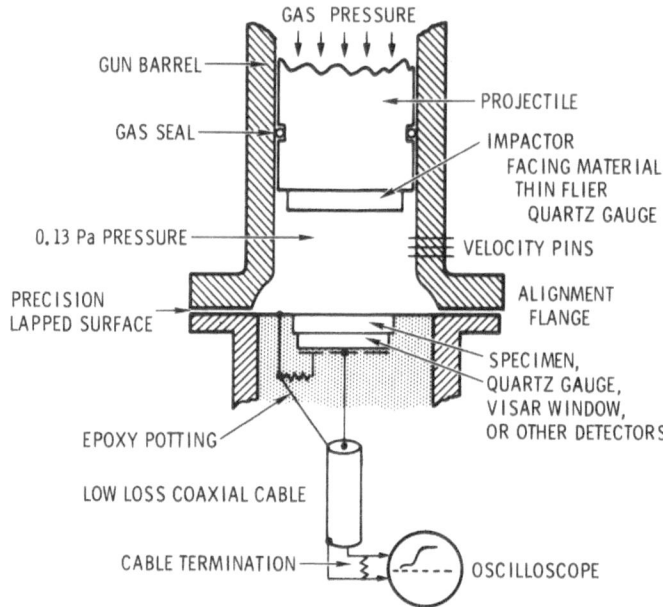

Fig. 3.5. The experimental arrangement used for a typical compressed gas gun is shown. The apparatus is designed to impact a selected impactor upon a target material with precision on the alignment of the impacting surfaces. Velocity at the impact surface can be measured to an accuracy and precision of 0.1%. This loading produces the most precisely known condition of all shock-compression events.

Wave Shaping and Shear Loading

When subjected to a step function loading, solid samples respond in one of the characteristic response modes described in Chap. 2. Often it is desired to investigate materials response to structured loading or even to shear-pulse loading. Both of these loadings can be achieved with the use of an intervening disk of a solid material placed between the loading and the sample.

The ramp of pressure to about 3 GPa observed in shock-loaded fused quartz has been used very effectively in acceleration-pulse loading studies of viscoelastic responses of polymers by Schuler and co-workers. The loading rates obtained at various thicknesses of fused quartz have been accurately characterized and data are summarized in Fig. 3.6. At higher peak pressures there are no precise standard materials to produce ramp loadings, but materials such as the ceramic "pyroceram" have been effectively employed. (See the description of the piezoelectric polymer in Chap. 5.)

Alternately (and showing the versatility of the impact technique), impactors can be designed to achieve structured loading. The "pillow" technique of Barker used a graded shock impedance to achieve a small amplitude shock followed by a slowly increasing pressure [88C04]. Materials synthesis studies

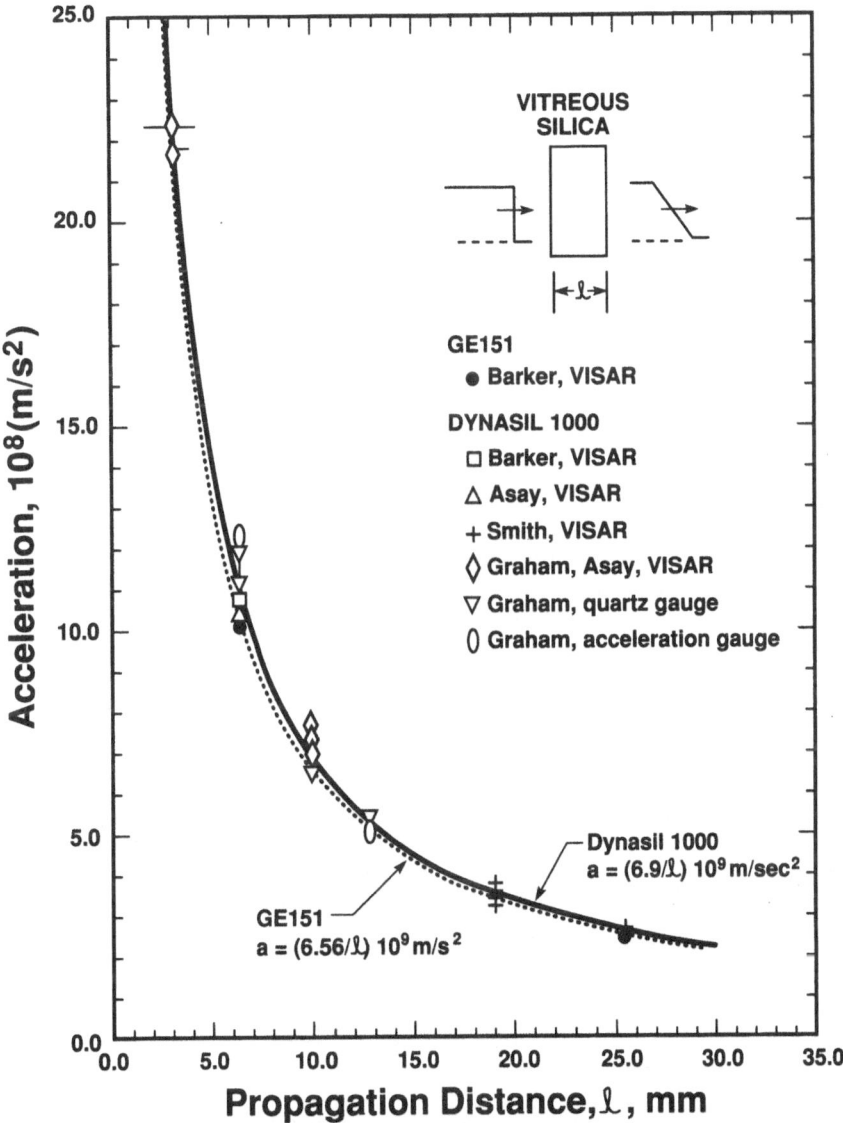

Fig. 3.6. The accelerations achieved at low pressure with waves transmitted through various thicknesses of fused quartz (GE 151 and Dynasil 1000) have been carefully studied and can serve as standard loadings (after Graham [79G02]). Recent data from Smith [92S01] also show the particle velocity limit for the linear acceleration to be 0.11 km s^{-1}.

in the Soviet Union have utilized layered impactors of solids of various thicknesses to achieve similar structured loadings [81A01].

Shear loading can be effectively achieved with the use of intervening high strength single crystals such as quartz, which, in a nonspecific direction, propagates quasilongitudinal and quasishear waves [80C01, 88C03]. Shear-pulse loading can also be achieved with the use of impactors and targets equally aligned at an angle to the axis of the projectile [80G04].

3.2 Measurement of Wave Profiles

Along with, and closely connected to, the developments in precise impact techniques is the development of methods to carry out time-resolved materials response measurements of stress or particle velocity wave profiles. With time resolutions approaching 1 ns, these devices have enabled study of mechanical responses not possible in the early period of the 1960s. The improved time-resolutions have resulted from direct measurement of stress or particle velocity, rather than from improved accuracy and resolution in measurement of position and time. In a continuation of this trend, capabilities are being developed to provide direct measurements of the rate-of-change of stress. With the ability to measure such a derivative function, detailed study of new phenomena and improved resolution and accuracy in descriptions of known rate-dependent phenomena seem possible.

Table 3.3 summarizes the history of the development of wave-profile measurement devices as they have developed since the early period. The devices are categorized in terms of the kinetic or kinematic parameter actually measured. From the table it should be noted that the earliest devices provided measurements of displacement versus time in either a discrete or continuous mode. The data from such measurements require differentiation to relate them to shock-conservation relations, and, unless constant pressures or particle velocities are involved, considerable accuracy can be lost in data processing.

The development of devices that provide a direct measure of stress or particle velocity led to observations of new rate-dependent mechanical responses and showed the power of such time-resolved measurements. The quartz gauge was the first of these devices with nanosecond time resolution, but its upper operating limit of 4 GPa limited its application. The development of the "VISAR" has had the most substantial impact on capabilities. VISAR systems, with time-resolution approaching 1 ns and the ability to work to pressures of 100 GPa, provide capabilities that have substantially altered the scientific descriptions of shock-compressed matter.

Over the 40-yr history of shock-compression science, numerous physical phenomena have been considered for use in detecting wave profiles. Few of the devices have actually been used for a significant and persistent study. Part of this history is connected to the difficulty in actually developing a credible

Table 3.3. Summary of wave profile instrumentation developments.

	Displacement vs time			
	Discrete	Continuous	Velocity- or stress-time	Acceleration-time
1945	pins	⋯	⋯	⋯
~	~	~	~	~
1955	flash gap	⋯	⋯	⋯
1956	⋯	⋯	⋯	⋯
1957	pins	capacitor	⋯	⋯
1958	⋯	⋯	⋯	⋯
1959	⋯	⋯	⋯	⋯
1960	⋯	⋯	*electromagnetic velocity capacitor	⋯
1961	⋯	optical image	*quartz piezoelectric	⋯
1962	⋯	⋯	*Manganin piezoresistant	⋯
1963	⋯	inclined mirror	optical lever	⋯
1964	pins	inclined resistor inclined prism	⋯	⋯
1965	pins	displacement interferometer	⋯	⋯
1966	⋯	⋯	⋯	⋯
1967	⋯	⋯	velocity interferometer	⋯
1968	pins	⋯	sapphire solid dielectric	⋯
1969	⋯	⋯	⋯	⋯
1970	⋯	electromagnetic	axisymmetric magnetic	⋯
1971	pins	⋯	⋯	⋯
1972	⋯	⋯	*VISAR	⋯
1973	⋯	⋯	LiNbO₃ piezoelectric	⋯
1974	⋯	⋯	carbon piezoresistant	⋯
1975	⋯	⋯	ytterbium piezoresistant	⋯
1976	pins	⋯	inclined electromagnetic	quartz piezoelectric
1977	pins	transverse interferometer	⋯	⋯
1978	⋯	⋯	shear interferometer	⋯
1979	⋯	⋯	⋯	⋯
1980	⋯	⋯	⋯	⋯
1981	pins	⋯	PVDF piezoelectric	PVDF piezoelectric
1982	⋯	bromoform	LiNbO₃ shear	⋯
1983	⋯	⋯	ORVIS	⋯
1984	pins	⋯	⋯	⋯

device. In Fig. 3.7 the development history of piezoelectric and piezoresistant stress gauges is summarized by noting the period of time from first publication of device study until the time of the last published study. Similar data are shown for electromagnetic and optical devices in Fig. 3.8. It is important to note that the record indicates that a period of at least 10 yr is required for device development, and the time can easily extend to 20 yr. The use of a particular device to carry out a significant materials study may take an addi-

EVOLUTION OF TECHNIQUES

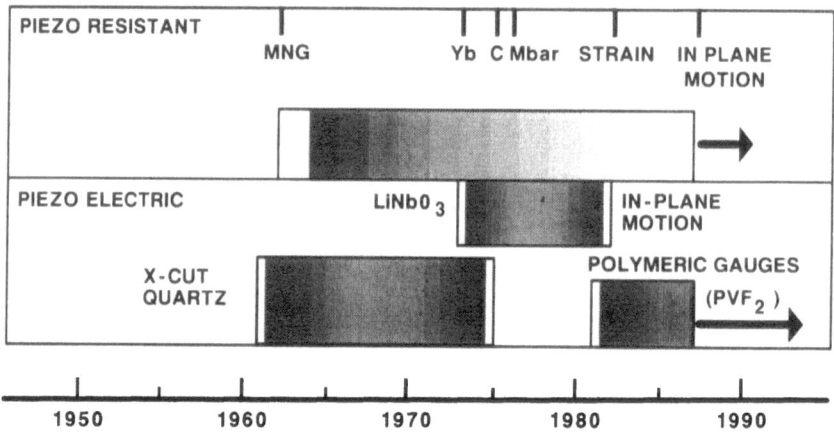

Fig. 3.7. The time periods over which significant developments are reported in the literature for the various piezoelectric and piezoresistant gauges are shown. It should be noted that the development period for development of such detectors is a minimum of 10 yr and possibly more typically 15 yr.

EVOLUTION OF TECHNIQUES

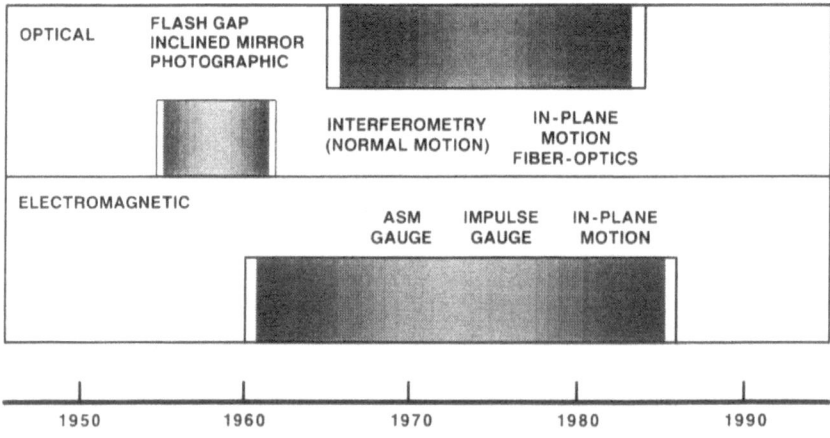

Fig. 3.8. The time periods over which significant developments are reported in the literature for the various electromagnetic and optical devices are shown. As for the electrical devices, the development periods are 10 yr or longer.

tional 5 or 10 yr for adaptation of the device to a particular materials study. The development of shock instrumentation is truly a monumental task! The level of effort and commitment requires unusually dedicated investigators and persistent funding.

3.3 Physical Categories of Detectors

Detectors that have been widely used for materials studies can be conveniently categorized by the physical phenomena utilized in the measurement as: piezoelectric, piezoresistant, electromagnetic, or optical. In each, the phenomena are capable of providing a signal in a time short compared to changes in characteristics of wave profiles. Given the desired time resolution of one nanosecond, the detectors must be "massless."

Piezoelectric stress gauges have active sensing materials of crystalline quartz and lithium niobate or of the piezoelectric polymer, polyvinylidene difluoride (PVDF). Piezoelectric copolymers are also currently being developed for gauge applications [90B03]. The piezoresistant stress gauges have active sensing elements of the metals Manganin, ytterbium, or carbon. Recent work incorporates constantan as a strain gauge to compensate for late-time strain effects in stress-gauge electrical leads.

Electromagnetic gauges include both the electromagnetic particle velocity and the axisymmetric magnetic gauges. Furthermore, an electromagnetic impulse gauge provides a measure of the integral of stress. These gauges utilize the signals produced by the motion of conducting sensors in insulating bodies. The recent development of fabrication techniques to provide multiple sensing elements that can be placed inside samples greatly improves the electromagnetic velocity gauge technique [89S01].

Optical devices or optical systems have provided most of the available strong shock data and were the primary tools used in the early shock-compression investigations. They are still the most widely used systems in fundamental studies of high explosives. The earliest systems, the "flash gap" and mirror systems on samples, provided discrete or continuous measurements of displacement versus time.

VISAR systems (*velocity interferometer system for any reflector*) are the most powerful sample response tool of modern shock-compression science. The capabilities have developed from the early concept of Barker of Sandia National Laboratories over a period of 15 yr into measurement systems for application over wide particle velocity ranges. The key development for higher velocity measurements is the development of "windows" placed in intimate contact with samples, which prevent problems associated with the interaction of high pressure shock waves with sample surfaces [86W01]. Recent development of the "fixed cavity" VISAR furnishes systems that are far more versatile [90S01]. Hemsing [79H01] has made significant revisions to the technology.

Although VISAR systems are expensive, and require significant commitment of personnel and data processing, the capabilities are unparalleled, and have truly changed descriptions of shock-compressed matter.

The various detectors can be categorized as utilizing physical property changes in sensors directly experiencing the shock compression or utilizing electromagnetic fields exterior to the shocked samples. In the first instance, credible gauge measurements rest on detailed study of the physical properties of the sensing materials over the range of shock conditions to be encountered; piezoelectric and piezoresistant gauges are examples of such gauges. In the second instance, the gauge or system utilizes optical or magnetic fields to probe samples. Changes caused by motion of the samples in the fields are sensed to provide the desired signals; optical and electromagnetic systems are examples of such systems. In general, a gauge requiring study of physical properties of sensors is more difficult to develop. In every case, however, careful and detailed study is required before a credible measurement can be accomplished.

3.4 Shock-Compression Gauges Cannot Be "Calibrated"

Perhaps the most widely misunderstood aspect of gauge development is the role of the controlled shock-compression experiment in the development process. It is often stated that the gauges are being "calibrated." In fact, it is not possible to calibrate a gauge that must be used over the wide range of conditions and over the wide range of wave profiles encountered and is destroyed in use. Only in special cases of "shocks" to fixed conditions is the response measured for a gauge in controlled experiments directly a suitable calibration. Even in the direct shock experiment, the controlled shock-compression experiment serves as a shock calibration only if the reproducibility of materials in the sensor is evaluated quantitatively and a persistent reproducible materials source is available.

In the more general case, the gauge must respond with fidelity in a known, quantitative manner to wave shapes with a variety of features. Responses measured under direct shock to fixed stresses cannot necessarily be extrapolated to the more general arbitrary wave shape. In a phenomenological sense, it must be assumed that the gauge response is independent of gauge-loading history. What is actually accomplished in the controlled "calibration" experiment is the identification of the physical effects encountered and quantification of responses, i.e., quantification of "the physics." With proper identification of the physical phenomena encountered, suitable models can be developed based on specific assumptions that can be evaluated over the range of shock-conditions encountered. The work reported in the following chapter on the piezoelectrics provides a specific example of such a gauge-development process.

The development of piezoresistant gauges provides a counterexample in that early work was based on empirical study of responses based on simple benign shock-compression concepts. More recently, Gupta has developed a physical basis for resistance changes in ytterbium and Manganin based on piezoresistive effects in metal lattices and contributions to resistance change due to shock-induced irreversible changes in defect concentration [87G01]. Piezoresistant stress gauges of standardized materials with known reproducibility are still not available. This feature is particularly critical as the time and effort required to develop a gauge requires such a concentration of effort over a number of years that the full characterization cannot be completed without the reproducible material source.

3.5 Advances in Measurement Technology

Studies of shock-compressed matter have progressed to a point for which detailed, sophisticated technology can probe mechanical responses in considerable detail. The detailed measurements now available appear to provide descriptions beyond that which can be predicted or fully interpreted on an established theoretical basis. As the conditions encountered are so unusual, a heavy reliance must be placed on the credibility of the experiments. Of particular importance is a recognition of the restricted view provided by a particular experiment, from both loading and sample response capabilities.

Given the advanced state of wave-profile detectors, it seems safe to recognize that the descriptions given by such an apparatus provide a necessary, but overly restricted, picture. As is described in later chapters of this book, shock-compressed matter displays a far more complex face when probed with electrical, magnetic, or optical techniques and when chemical changes are considered. It appears that realistic descriptive pictures require probing matter with a full array of modern probes. The "recovery" experiment in which samples are preserved for post-shock analysis appears critical for the development of a more detailed defective solid scientific description.

For mechanical wave measurements, notice should be taken of the advances in technology. It is particularly notable that the major advances in materials description have not resulted so much from improved resolution in measurement of displacement and/or time, but in direct measurements of the derivative functions of acceleration, stress rate, and density rate as called for in the theory of structured wave propagation. Future developments, such as can be anticipated with piezoelectric polymers, in which direct measurements are made of rate-of-change of stress or particle velocity should lead to the observation of recognized mechanical effects in more detail, and perhaps the identification of new mechanical phenomena.

PART III

Physical Properties of
Shock-Compressed Solids

CHAPTER 4

Physical Properties Under Elastic Shock Compression

In this chapter: nonlinear piezoelectric and dielectric behavior; shock-induced electrical conductance; semiconductors; elastic physical properties.

Conventional physical descriptions of materials in the solid state are concerned with solids in which properties are controlled or substantially influenced by the crystal lattice. When defects are treated in typical solid state studies, they are considered to modify and cause local perturbations to bonding controlled by lattice properties. In these cases, defect concentrations are typically low and usually characterized as either point, linear, or higher-order defects, which are seldom encountered together.

When solids are subjected to shock compression above their Hugoniot elastic limits, the shear stresses resulting from the uniaxial strain exceed the strength of the solid, and extensive plastic deformation occurs on a submicrosecond time scale. The rapid motion of defects leads to high concentrations of both point and higher-order defects. These defect states are qualitatively and quantitatively different from those encountered in other mechanical loading environments. Given the complications introduced by unusually high concentrations of defects in highly compressed lattices, it is of interest to consider shock-compressed solids within their purely elastic ranges to investigate how typical solid properties change at large elastic compression. Based on such studies, the elastic strain portion of higher pressure shock compression can then be evaluated in the more complex states. It should be noted that within the benign shock-compression paradigm the problem is considered to consist of only the compressed lattice; the possible complications of saturation levels of defects in highly compressed lattices are not explicitly considered.

In this chapter physical properties of solids at finite strain within their purely elastic ranges will be investigated. Although the strain levels of a few percent are small relative to the total compressions of typical shock-compression studies, they are large compared to those typically encountered in higher-order elastic property investigations.

4.1 Nonlinear Piezoelectric Properties

Favorable electrical properties and large Hugoniot elastic limits, combined with ready availability, have led to the widespread use of quartz and lithium niobate crystals as time-resolved stress gauges in shock-compression experiments. The importance of this application has motivated sufficiently quantitative studies that their piezoelectric, dielectric, and elastic properties have been determined in detail throughout the elastic range. The principal distinctive results from these investigations are determinations of second-order piezoelectric, higher-order piezoelectric, dielectric, and elastic constants, and investigation of unusual shock-induced dielectric breakdown phenomena. Much of this work is summarized in Davison and Graham [79D01] and Graham and Reed [78G02]. In this section we revert to the continuum elasticity sign convention of *positive values for tensile stress and strain*.

Studies of the response of piezoelectric solids to elastic shock compression are part of a larger question of nonlinear piezoelectric response. Although this problem is of considerable interest in connection with microwave acoustic phenomena, there are few quantitative data on nonlinear piezoelectric constants. Order-of-magnitude estimates based on ultrasonic investigations have been given for lithium niobate [75K02], while quantitative values are reported for quartz by Hruska [78H03], who was the first to detect a nonlinear piezoelectric effect [61H02]. Pressure derivatives of hydrostatic piezoelectric constants have been accurately measured to 2.6 GPa for lithium niobate and lithium tantalate [76G04].

Elastic Dielectric Theory

When a stress pulse is propagated into a piezoelectric solid, the resulting strain produces a local polarization through the direct piezoelectric effect. This polarization causes electric fields to develop in the space between the electrodes and an associated current to flow in an external circuit connecting the electrodes. The fields cause secondary stresses to develop in the sample through the indirect piezoelectric effect. The magnitude and distribution of the electric fields depend on the form of the stress pulse and the character of the external circuit, while details of the secondary stress depend on the electromechanical coupling coefficients and the mechanical boundary conditions at the electrodes. This coupling between electrical and mechanical effects, which is a fundamental characteristic of piezoelectric materials, considerably complicates analysis of their response to mechanical loads. In the present case nonlinear effects cause further complication. In fact, no fully coupled, closed-form solution for a nonlinear dynamic piezoelectric response has been developed, although solutions have been obtained numerically [76C02, 77L01]. Fortunately, electromechanical coupling is often weak and this fact can be taken advantage of to obtain approximate solutions that are accurate to within a few percent.

Constitutive Relations

Piezoelectric solids are characterized by constitutive relations among the stress t, strain η, entropy s, electric field E, and electric displacement D. When uncoupled solutions are sought, it is convenient to express t and D as functions of η, E, and s. The formulation of nonlinear piezoelectric constitutive relations has been considered by numerous authors (see the list cited in [77G06]), but there is no generally accepted form or notation. With some modification in notation, we adopt the definitions of thermodynamic potentials developed by Thurston [74T01]. This leads to the following constitutive relations:

$$t_{rs} = \frac{\rho}{\rho_R} F_{ri} F_{sj} (C^E_{ijkl}\eta_{kl} - e_{kij}E_k + \tfrac{1}{2}C^E_{ijklmn}\eta_{kl}\eta_{mn}$$

$$+ \tfrac{1}{6}C^E_{ijklmnpq}\eta_{kl}\eta_{mn}\eta_{pq} - \tfrac{1}{2}f_{klij}E_kE_l - \tfrac{1}{2}\varepsilon_{ijklm}E_m\eta_{kl}), \qquad (4.1)$$

$$D_i = e_{ijk}\eta_{jk} + \varepsilon^\eta_{ij}E_j + \tfrac{1}{2}e_{ijklm}\eta_{jk}\eta_{lm} + \tfrac{1}{2}f_{ijkl}E_j\eta_{kl} + \tfrac{1}{2}\varepsilon^\eta_{ijk}E_jE_k,$$

in the independent variables η, E, and s (the coefficient tensors are functions of s). In these equations, the tensor components C^E_{ijk}, D^E_{ijklmn}, and $C^E_{ijklmnpq}$ are second-, third-, and fourth-order elastic stiffness coefficients at constant field, e_{kij} and e_{kijlm} are second- and third-order piezoelectric stress constants, ε^η_{ij} and ε^η_{ijk} are second- and third-order dielectric permittivities at constant strain, and f_{ijkl} is the electrostrictive coefficient. In applying these relations, the principal stress and strain components are positive in tension. Small pyroelectric contributions to electric displacement due to isentropic heating of certain crystals are treated elsewhere [77G06]. The contribution $(\tfrac{1}{2})\varepsilon^\eta_{ijk}E_jE_k$ to the electric displacement of x-cut quartz is extremely small and is neglected in analysis of both quartz and lithium niobate.

For many problems it is convenient to separate the piezoelectric (i.e., strain induced) polarization P^η from electric-field-induced polarizations by defining $D = P^\eta + \varepsilon E$, where ε is the permittivity tensor. For uniaxial strain and electric field along the 1 axis, when the material is described by Eq. (4.1) with the E^2 term omitted,

$$P_1\eta = [e_{11} + (1/2)e_{111}\eta_1]\eta_1,$$

$$\varepsilon_{11} = \varepsilon^\eta_{11} + (1/2)f_{111}\eta_{11}. \qquad (4.2)$$

Mechanical wave-propagation problems are analyzed on the basis of quasi-static electromagnetic conditions. This is an excellent approximation since the electromagnetic wavespeed greatly exceeds the mechanical wavespeed and the particle velocity is typically only about one-tenth of the mechanical wavespeed. (See the discussion by Thurston [74T01].)

Configurations of interest are those using disk-shaped samples cut from crystals in orientations that permit plane waves of uniaxial strain to propagate through their thickness when a uniform load is applied to their face. When the diameter of the disk is sufficiently large in comparison to its thick-

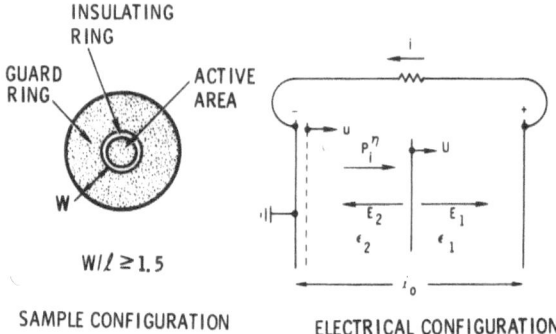

Fig. 4.1. As shown on the left, the configuration of conducting plates on the flat faces of piezoelectric disks produces one-dimensional strain and electric field conditions with a guard-ring arrangement. On the right, the typical electrostatic conditions are shown. The axis through the thickness of the disk is chosen as the x axis.

ness, and a suitable guard-ring electrode configuration is used as shown in Fig. 4.1, the fields throughout the inner region of the sample will be normal to the faces of the disk [65G01]. In such a one-dimensional strain and electric field configuration, the current pulse depends only on the area of the charge-collecting electrode and the thickness between electrodes. Thus, all current pulses at the same stress are identical when normalized to these geometric variables.

In the absence of free charge in the disk, the electric displacement will be independent of position, although it will vary with time: $D = [D(t), 0, 0]$. The current induced in the external circuit is attributable to changes in electric displacement within the disk and is given by

$$i(t) = A \frac{dD(t)}{dt}, \tag{4.3}$$

where A is the area of the charge-collecting electrode on the face of the disk. The voltage across the disk is the integral of the electric field in the space between the electrodes,

$$V(t) = \int_{\chi(0,t)}^{\chi(L,t)} E(\chi, t) \, d\chi. \tag{4.4}$$

When the electrodes are connected by a short circuit, the voltage is zero but only the average field need vanish; local values are often quite large.

Piezoelectric Response: Uncoupled Short-Circuit Solution

The "uncoupled" response of a piezoelectric sample to elastic shock compression is determined on the assumption that the mechanical response of the material is independent of any electric fields that may be present. In this approximation, a steady shock is introduced into material at rest while the part behind the shock will be uniformly compressed to a strain $S_1 = -u/U$, where u is the particle velocity of the compressed material. The compression will produce a piezoelectric polarization $P^\eta = (P^\eta, 0, 0)$ in the compressed material and will cause a change in the permittivity component in the x direction from some value ε_1 to a new value ε_2 that depends on the strain according to Eq. (4.2). From Eq. (4.2)–(4.4), it can be shown that, when the electrodes are connected by a short circuit $[V(t) = 0]$,

$$\frac{i(t)L}{P^\eta AU} = \frac{\alpha(1 - u/U)}{[(1 - u/U)(t/t_0) + \alpha(1 - t/t_0)]^2}, \qquad 0 < t < t_0, \qquad (4.5)$$

where x is the axial direction, L is the original thickness of the sample, $t = 0$ at the instant of introduction of the shock into the sample, $t_0 = L/U$ is the transit time of the shock through the sample, and $\alpha = \varepsilon_2/\varepsilon_1$. From Eq. (4.2), we see that the magnitude of the piezoelectric polarization is directly dependent on the magnitude of the strain, but Eq. (4.5) shows that this dependency does not affect the current history for the case considered. For both quartz and lithium niobate, $u/U < 0.04$ and $1 \leq \alpha < 1.01$ in the elastic range. With these restrictions, Eq. (4.5) indicates that the current history is a step function to close approximation. Deviations from this ideal shape shown in the equation occur as a geometrical result of large compressions or when the permittivity changes significantly upon compression.

The current immediately after impact, $i(0+)$, is calculated from the limiting case

$$\frac{i(0+)L}{AU} = \frac{1 - (u/U)}{\alpha} P^\eta \qquad (4.6)$$

of Eq. (4.5). This relation holds universally for shock-induced polarization effects and is useful for evaluation of various shock-induced polarization phenomena.

In the low-signal limit in which nonlinearities in material behavior are negligible and $u/U \ll 1$ the analysis given above can easily be extended to stress pulses of arbitrary form, with the result [65G01]

$$\frac{i(t)L}{AU} = \frac{e_{11}}{C_{11}^E} t_{11}(0, t), \qquad 0 < t < t_0, \qquad (4.7)$$

which indicates that the current history is proportional to the history of stress at the input electrode $t_{11}(0, t)$. This relation, which is also followed to a close approximation at larger strains if e_{11}/C_{11}^E is replaced by an experimentally determined, stress-dependent coefficient, forms the basis for the widely used current-mode piezoelectric gauges [65G01, 75G04].

Electric Fields

Shock compression of piezoelectric solids, even under short-circuit conditions, causes large electric fields of varying amplitude and polarity within the material. In the uncoupled approximation to the solution of the short-circuit problem, the field is easily determined from the condition on uniformity of electric displacement, from Eq. (4.7), and the expression $D = P^\eta + \varepsilon E$,

$$E = \frac{P^\eta/\varepsilon_{11}}{(1 - u/U)(t/t_0) + \alpha(1 - t/t_0)} \begin{cases} (1 - u/U)(t/t_0); & \text{for } \chi > Ut \\ 1 - t/t_0; & \text{for } \chi < Ut \end{cases}, \quad (4.8)$$

for $0 < t < t_0$. The magnitude of the field in each region varies between zero and a maximum value of P_η/ε_{11} during passage of the shock through the sample. Since u/U is small and α near unity for most cases of interest, the field at a given point varies approximately linearly with time except for a discontinuity when the shock passes. When a 2 GPa shock passes through x-cut quartz, the maximum field strength is about 10^8 V m^{-1}, while a maximum field of 3×10^7 V m^{-1} is realized in z-cut lithium niobate subjected to a stress slightly in excess of 1 GPa. Fields of these magnitudes are of concern in that they are about one-tenth of breakdown strength values at atmospheric pressure. From this analysis we see that shock-compressed piezoelectric materials are subjected to the simultaneous effects of high stress and high electric field. Such fields are found to be crucial in activating mechanisms of shock-induced conduction.

Weakly and Fully Coupled Solutions

Uncoupled solutions for current and electric field give simple and explicit descriptions of the response of piezoelectric solids to shock compression, but the neglect of the influence of the electric field on mechanical behavior (i.e., the electromechanical coupling effects) is a troublesome inconsistency. A first step toward an improved solution is a weak-coupling approximation in which it is recognized that the effects of coupling may be relatively small in certain materials and it is assumed that electromechanical effects can be treated as a perturbation on the uncoupled solution.

The contribution to the stress from electromechanical coupling is readily estimated from the constitutive relation [Eq. (4.2)]. Under conditions of uniaxial strain and field, and for an open circuit, we find that the elastic stiffness is increased by the multiplying factor $(1 + K^2)$ where $K^2 = e_{11}^2/(\varepsilon_{11}^\eta C_{11}^E)$ the square of the electromechanical coupling factor for uniaxial strain, is a measure of the stiffening effect of the electric field. Values of K^2 for various materials are for x-cut quartz, 0.0008, for z-cut lithium niobate, 0.055: for y-cut lithium niobate, 0.074; for barium titanate ceramic, 0.5; and for PZT-5H ceramic, 0.75. These examples show that electromechanical coupling effects can be expected to vary from barely detectable to quite substantial.

Stuetzer [67S02, 67S03] and Thurston [74T01] have determined the coupled response of linear piezoelectrics to step loading, while both Lysne [72L02] and Thurston [74T01] have obtained solutions for the corresponding problem for weakly coupled nonlinear piezoelectrics. In each case the short-circuit current exhibits the same initial jump as for the uncoupled solution, with the current at later times becoming greater in the coupled than in the uncoupled case by an amount that depends on the electromechanical coupling factor for the material and the mechanical boundary conditions to which the sample disk is subjected.

Chen et al. [76C02] and Lawrence and Davison [77L01] have placed the fully coupled nonlinear theory of uniaxial piezoelectric response in a form that is convenient for numerical solution of problems and have simulated a number of experiments in terms of this theory. An example of the results obtained is given below.

From a constitutive relation of the form $t = t(D, \eta)$ (here t is stress not time) it can be readily shown that, since there is no change in electric displacement in an open-circuit, thick-sample configuration, there are no secondary stresses due to electromechanical coupling. Nevertheless, the wavespeed is that of a piezoelectrically stiffened wave.

Experimental

The piezoelectric behavior of both quartz and lithium niobate has been studied in a series of careful, systematic investigations. (See Graham and co-workers [65G01, 70I01, 75G04].) The experimental arrangement is shown Fig. 4.2. The impactor, preferably the same material as the piezoelectric sample (but perhaps another standard material), is accelerated to a preselected

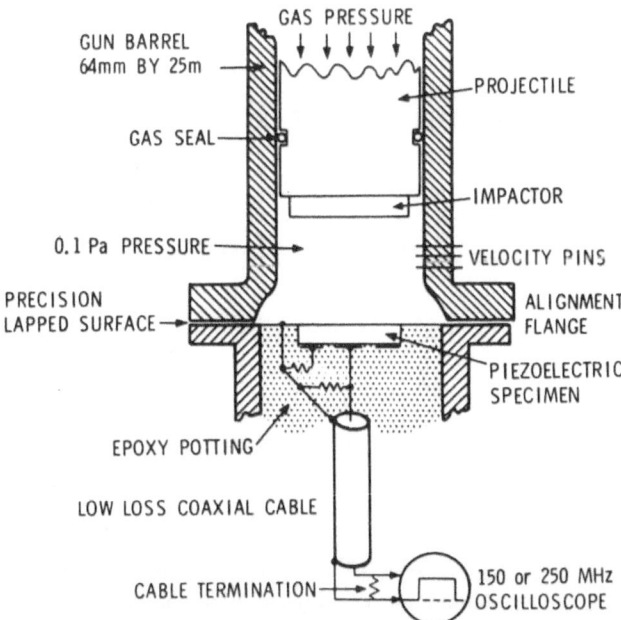

Fig. 4.2. The technique used to study the piezoelectric behavior of the crystals quartz and lithium niobate used controlled, precise impact loading. The impact velocity can be measured to an accuracy of 0.1%, leading to the most precisely known condition in shock-compression science (after Davison and Graham [79D01]).

velocity and impacted, in vacuum, upon the sample. Measured quantities include the impactor velocity immediately prior to impact and the short-circuited current pulse produced during the passage of the shock through the sample. Based on Eq. (4.6), each experiment yields a value for piezoelectric polarization at a given strain; a collection of such data over a wide range of strain permits the linear and nonlinear piezoelectric constants to be deter-mined. The current-pulse amplitude can be measured to an accuracy of $\pm 1\%$ and, since the impact velocities from which strains are computed are known to $\pm 0.1\%$, overall accuracies are excellent. Error in shock velocity does not cause error in determination of the piezoelectric stress constants.

Typical current pulses observed for x-cut quartz, z-cut lithium niobate, and y-cut lithium niobate are shown in Fig. 4.3. Following a sharp rise in current to an initial value (the initial rise time is due to "tilt," misalignment of the impacting surfaces), the wave shapes show either modest increases in current during the wave transit time for quartz and z-cut lithium niobate

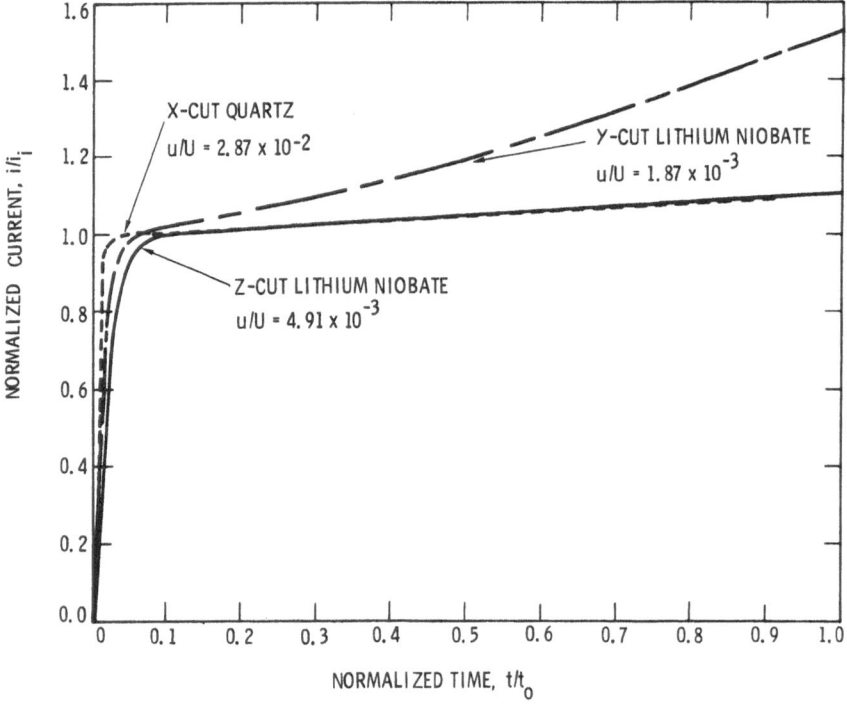

Fig. 4.3. Typical normalized piezoelectric current-versus-time responses are compared for x-cut quartz, z-cut lithium niobate, and y-cut lithium niobate. The y-cut response is distorted in time due to propagation of both longitudinal and shear components. In the other crystals, the increases of current in time can be described with finite strain, dielectric constant change, and electromechanical coupling as predicted by theory (after Davison and Graham [79D01]).

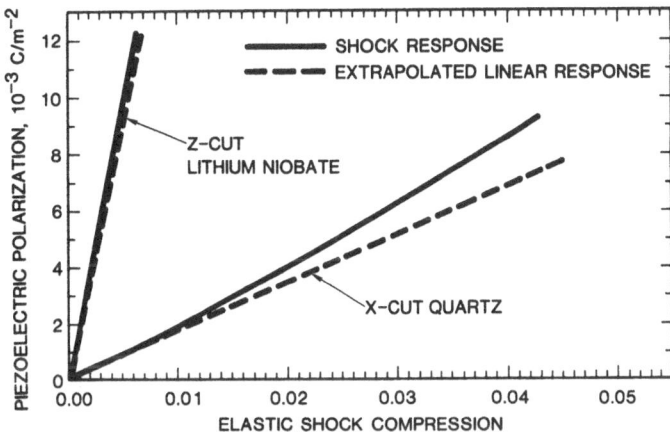

Fig. 4.4. The piezoelectric charge produced by elastic strain in x-cut quartz and z-cut lithium niobate is well represented by a quadratic relationship without a need for fourth-order contributions.

samples in agreement with the theory, or large increases in current for y-cut lithium niobate samples. Given our previous discussions of electromechanical coupling, it can be determined that the large increase in current with time in y-cut lithium niobate is an indication of the strong influence of such coupling. It should be noted that current pulse distortions are also significant from samples without proper guard rings to ensure one-dimensional conditions. Subtle, but significant, distortions in current pulses are obtained from samples in the "shorted guard-ring" configuration [75G04].

The measured relationships between piezoelectric polarization and strain for x-cut quartz and z-cut lithium niobate are found to be well fit by a quadratic relation as shown in Fig. 4.4. In both materials a significant non-linear piezoelectric effect is indicated. The effect in lithium niobate is particularly notable because the measurements are limited to much smaller strains than those to which quartz can be subjected. The quadratic polynomial fits are used to determine the second- and third-order piezoelectric constants and are summarized in Table 4.1. Elastic constants determined in these investigations were shown in Chap. 2.

In the use of piezoelectric crystals for stress-pulse measurements, it is convenient to describe the current pulse in terms of the initial current jump i_i for step loading based on Eq. (4.7) analogous to e_{11}/c_{11}. The piezoelectric current coefficient k thus defined can then be used as a "calibration factor" when combined with a representation of the increase in current during wave transit time. A summary of these representations is shown in Table 4.2, which can be used for gauge applications.

In the case of x-cut quartz there is excellent agreement between second-order constants determined in the shock-compression studies and ultrasonic

Table 4.1. Second- and third-order piezoelectric
constants (after Davison and Graham [79D01]).

Sample	$e_{ij}(\mathrm{C/m^2})$	$e_{ijk}(\mathrm{C/m^2})$	e_{ijk}/e_{ik}
Lithium niobate			
$i = j = k = 3$	1.80	21	-11
$i = i = k = 2$	2.37	21	$+9$
$i = l, j = 5$	3.83	$\cdots$	$\cdots$
rotated cut	4.65	0	$\cdots$
Quartz			
$i = j = k = 1$	0.171	-2.62	-15.3

Table 4.2. Calibration factors for quartz and lithium niobate gauges.

Material	Relative sensitivity	Output nonlinearity (% per GPa)	Distortion*	Stress limit (GPa)
x-cut quartz	1.0***	$+4.8$	1.04	4.0
z-cut lithium niobate	4.6	$+2.5$	1.08	0.8–1.4**
y-cut lithium niobate	6.5	-1.3	1.51	1.8
36° rotated lithium niobate	11.8	0	1.51	0.8

See the papers summarized in Graham and Reed [78G02].
* Distortion is the ratio of final current level to initial current level during transit of the stress pulse from a step-function input.
** Different limiting stresses are observed depending on gauge thickness. The higher limits are characteristic of thicker gauges.
*** $k(\sigma) = (2.00 + 0.097\sigma) \times 10^{-3}$ C/cm^2 GPa^{-1} for x-cut quartz. Wave speed for quartz is 5.72 km s^{-1}; for z-cut lithium niobate 7.33 km s^{-1}; y-cut lithium niobate 6.9 km s^{-1}; rotated cut lithium niobate 7.35 km s^{-1}.

investigations. The third-order piezoelectric constants for quartz determined from the impact experiments are determined far more accurately than those determined from ultrasonic studies.

For lithium niobate, the second-order constants for shock and ultrasonic studies are in good agreement except for the e_{33} constant, which is observed to be higher in shock studies than in ultrasonic investigations [77G06]. This discrepancy is thought to be due to the use of incompletely poled material for the earlier ultrasonic work. The third-order constants for lithium niobate are not determined as accurately as those for x-cut quartz due to the relatively low strains which can be applied before the onset of shock-induced dielectric breakdown (discussed later in this chapter). Nevertheless, the errors of these constants are lower than the order-of-magnitude estimates obtainable by ultrasonic means.

Lithium niobate is strongly ferroelectric, yet the material behavior under elastic shock loading is apparently fully described by nonlinear piezoelec-

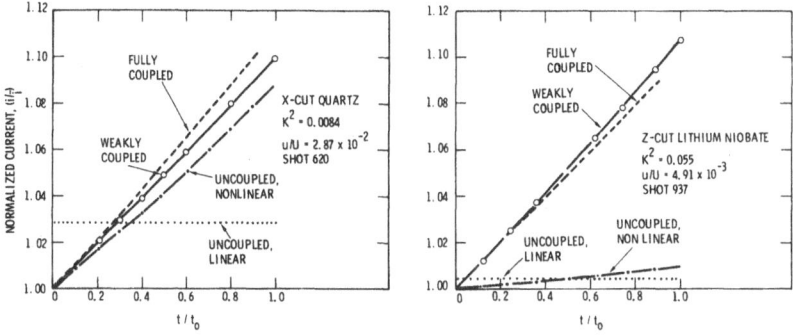

Fig. 4.5. The degree of approximation for the increase of current in time for uncoupled and weakly coupled solutions for impact-loaded, x-cut quartz and z-cut lithium niobate is shown by comparison to the numerically predicted, fully coupled case. In the figure, the initial current is set to the value of 1.0 at the measured value (after Davison and Graham [79D01]).

tricity [77G06]. This is not unreasonable since it is well known that domain realignment with field occurs only in the vicinity of the Curie temperature of 1475 K.

The ratio of third- to second-order piezoelectric constants has also been determined for x-cut quartz with the acceleration pulse loading method [77G05]. Two experiments yielded values for e_{111}/e_{11} of 15.0 and 16.6 compared to the ratio of 15.3 [72G03] determined from the fit to the 25 shock loading experiments.

The determination of piezoelectric constants from current pulses is based on interpretation of wave shapes in the weak-coupling approximation. It is of interest to use the wave shapes to evaluate the degree of approximation involved in the various models of piezoelectric response. Such an evaluation is shown in Fig. 4.5, in which normalized current-time wave forms calculated from various models are shown for x-cut quartz and z-cut lithium niobate. In both cases the differences between the fully coupled and weakly coupled solutions are observed to be about 1%, which is within the accuracy limits of the calculations. Hence, for both quartz and lithium niobate, weakly coupled solutions appear adequate for interpretation of observed current-time waveforms. On the other hand, the adequacy of the uncoupled solution is significantly different for the two materials. For x-cut quartz the maximum error of about 1%–1.5% for the nonlinear-uncoupled solution is suitable for all but the most precise interpretation. For z-cut lithium niobate the maximum error of about 8% for the nonlinear-uncoupled solution is greater than that considered acceptable for most cases. The linear-uncoupled solution is seriously in error in each case as it neglects both strain and coupling.

A unique electrical-to-mechanical coupling effect called "piezoelectric rate coupling" has been predicted to occur in the neighborhood of a shock in nonlinear piezoelectric solids [75G05]. The effect appears as a strain gradient in the presence of an electric field rate. The strain gradient apparently persists for only a few nanoseconds or is of such small magnitude that it cannot be observed in careful measurements with a VISAR system since such un-published measurements by Graham and Asay found no evidence for the expected coupling.

Discussion of Experimental Measurements

The direct nature of the piezoelectric polarization measurements under shock have permitted longitudinal second- and third-order piezoelectric constant determinations for x-cut quartz and z-cut lithium niobate which are probably the most accurate determined by any method. The precision and detail with which this problem has been studied far exceeds that for other electrical effects under shock loading. Unfortunately, a successful investigation requires numerous samples, and conditions of inelastic deformation and shock-induced conduction must be avoided; such conditions cannot be achieved in a wide class of piezoelectric materials. There is hope that the acceleration loading technique, which requires only a single sample whose properties are determined at low strain, can make higher-order piezoelectric constant measurements applicable to a broader class of piezoelectrics. The use of transverse shock loading and detection should, in principle, make it possible to determine a full set of higher-order constants. Gupta [84G04] has reported the first measurement of the piezoelectric response to a shear wave in impacted lithium niobate.

Several structural theories of piezoelectricity [72M01, 72M02, 72A05, 74H03] have been proposed but apparently none have been found entirely satisfactory, and nonlinear piezoelectricity is not explicitly treated. With such limited second-order theories, physical interpretations of higher-order piezoelectric constants are speculative, but such speculations may help to place some constraints on an acceptable piezoelectric theory.

Perhaps the most far-reaching observation from the present measurements under large uniaxial strain and from similar measurements under large hydrostatic pressures [76G04] is that fourth-order piezoelectric effects are negligibly small even though they should be readily detectable at the strain encountered. These observations are in sharp contrast to elastic effects in which fourth-order contributions are readily observed. The details of inter-atomic interactions which lead to undetectable fourth-order effects cannot be directly determined but it is noteworthy that such behavior can be described in terms of two competing interactions, each with a constant but different strain derivative over a large range of strain. Martin's theory of piezoelectricity applied to zinc blende crystals [72M01, 72M02] is explicitly expressed in terms of two opposing interactions, the rigid ion dipole interaction and a quadrupolar interaction which involves charge redistribution. Hanson and co-workers [77H04] have used Martin's theory to interpret their nonlinear piezoelectric constant data for CuCl under hydrostatic pressure.

A published collection of selected reprints concerning piezoelectricity under shock loading and related piezoelectric gauges is available as a convenient source for work in the area of shock-compressed elastic piezoelectrics [78G02].

Linear Piezoelectric Pulse Analysis

Electrostatic models reported earlier in this chapter represent special cases of piezoelectric polarization pulses propagating through samples in response to shock loading. In analysis of the data, time dependent stresses will be encountered that will change the piezoelectric pulse shape. In that case a more flexible model is needed to relate the variables for a piezoelectric polarization pulse of arbitrary profile propagating through the sample. Although a completely general theory appears too difficult for immediate solution, a linearized theory that treats small deformations in a linearly elastic dielectric of constant permittivity can serve as a useful approximation. In the present model, the piezoelectric effect may be nonlinear; however, it is again convenient to incorporate the weak-coupling approximation.

Changes in polarization may be caused by either the input stress profile or a relaxation of stress in the piezoelectric material. The mechanical relaxation is obviously inelastic but the present model should serve as an approximation to the inelastic behavior. Internal conduction is not treated in the theory; nevertheless, if electrical relaxations in current due to conduction are not large, an approximate solution is obtained. The analysis is particularly useful for determining the signs and magnitudes of the electric fields so that threshold conditions for conduction can be established.

It has previously been shown that, within the approximation above, an arbitrary stress profile applied to a piezoelectric disk produces a current

$$i(t) = (fAU/l)\sigma(0, t), \quad 0 < t < t_0, \tag{4.9}$$

where f is a coefficient that relates polarization to stress and $\sigma(0, t)$ is the stress at $\chi = 0$, the input electrode. More generally, the same relationship can be expressed as

$$i(t) = (AU/l)P(0, t), \tag{4.10}$$

where $P(0, t)$ is the polarization at the input electrode. Equation (3.13) provides a relationship needed to relate an arbitrary polarization profile to a measured or assumed current pulse.

The polarization moves with the stress profile at a velocity U; hence, once the profile is specified at an initial location the distribution of polarization at another location can be determined as

$$P(\chi, t) = P(0, t - \chi/U), \quad \chi/U < t < t_0. \tag{4.11}$$

If no conduction occurs, the displacement at all locations within the disk has a uniform value

$$D(t) = \frac{1}{A} \int i(t)\, dt. \tag{4.12}$$

The displacement is the sum of the polarization and εE; hence, there is sufficient information to specify the electric field. It follows that

$$\varepsilon E(\chi, t) = P(\chi, t) - D(t), \tag{4.13}$$

where the relationships for $D(t)$ and $P(\chi, t)$ are determined from Eqs. (4.12) and (4.13). Frequently the field at the $\chi = 0$ location is all that is required. In that case the relation

$$\varepsilon E(0, t) = + \frac{1}{AU} i(t) - \frac{1}{A} \int i(t)\, dt \tag{4.14}$$

allows a direct solution for the field from the measured current profile. The sign of the electric field from Eq. (4.13) is designated such that a positive field corresponds to that initially produced in the stressed region of a plus-x orientation sample. Previous investigations within the elastic range have shown that a negative value for a field of the correct magnitude would be expected to cause the catastrophic change in conductivity associated with dielectric breakdown.

Equation (4.14) is depicted graphically in Fig. 4.6. The polarization function $P(0, t)$ represents a relaxation of polarization at the input electrode. The $D(t)$ function represents the integral of the corresponding current pulse. In Fig. 4.6 the field at any location may be determined as the difference between $P(t)$ and $D(t)$. Early in time the field is noted to decrease as the polarization decreases and the displacement increases. At time (2) the field is zero and in

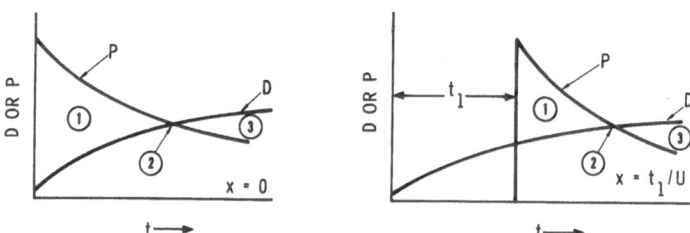

Fig. 4.6. Piezoelectric pulse diagrams can be used to obtain explicit representations of the time dependent electric fields in piezoelectric substances. The magnitudes and orientations of these electric fields are critical to development of shock-induced conduction. As an example, the diagram on the left shows the polarization and displacement relations for a location at the input electrode. The same functions for a location within the crystal is shown on the right (after Davison and Graham [79D01]).

region (3) the field is negative. The example shows that within the limitations of the assumptions both the signs and magnitudes of the electric fields can be determined from a known polarization function and that a reduction in polarization will lead to a sign reversal of the electric field.

The linear piezoelectric model can be used to demonstrate that the magnitude of the electric field encountered for a given polarization function is a sensitive function of the thickness of the sample. This behavior can be demonstrated by noting that the electric displacement at a given time is inversely proportional to the thickness. Thus, the thickness of the sample is an important variable for investigating effects such as conductivity that depend upon the magnitude of the electric field. Conversely, various input stress wave shapes can be used to cause various field distributions at fixed thicknesses.

4.2 Normal Dielectrics

Nonlinear properties of normal dielectrics can be studied in the elastic regime by the method of shock compression in much the same way nonlinear piezoelectric properties have been studied. In the earlier analysis it was shown that the shape of the current pulse delivered to a short circuit by a shock-compressed piezoelectric disk was influenced by strain-induced changes in permittivity. When a normal dielectric disk is biased by an electric field and is subjected to shock compression, a current pulse is also delivered into an external circuit. In the short-circuit approximation, the amplitude of this current pulse provides a direct measure of the shock-induced change in permittivity of the dielectric.

A normal dielectric may be characterized by Eq. (4.1) with the piezoelectric terms deleted. For an isotropic dielectric subject to uniaxial strain and a collinear electric field this equation takes the form

$$D_1 = [\varepsilon^\eta_{11} + 1/2(\varepsilon^\eta_{111}E_1) + 1/2(f_{111}\eta_1)]E_1, \quad D_2 = D_3 = 0. \quad (4.15)$$

Neglecting the small effect of electrostrictive coupling on mechanical behavior, we see from Eq. (3.4) that shock propagation is not influenced by electrical effects. Under this approximation, a steady shock propagated into the material will divide its thickness into two regions of uniform strain that can be analyzed in the same manner as for the piezoelectric response. In the absence of free charge, Eqs. (4.3) and (4.4) applied to an elastic disk of thickness L having an electrode of area A and subject to a potential V yield the relation [68G05]:

$$\frac{i(t)l^2}{AVU\varepsilon^+} = \left[\frac{u}{U} + \frac{\Delta\varepsilon}{\varepsilon^+}\left(1 + \frac{u}{U}\right) + \left(\frac{\Delta\varepsilon}{\varepsilon^+}\right)^2\right]\left[1 - \frac{ut}{L} + \frac{\Delta\varepsilon}{\varepsilon^+}\left(1 - \frac{t}{t_0}\right)\right]^{-2} \quad (4.16)$$

for the time interval $0 < t < t_0 = l/U$ after impact. In this relation ε^+ is the permittivity of the uncompressed material, i.e., the coefficient of E_1 in Eq.

(4.16) evaluated at $\eta_1 = 0$, $E_1 = V/L$, and $\Delta\varepsilon$ is the change in this coefficient that occurs with passage of the wave. The change in permittivity, which is proportional to the electrostrictive constant f_{111}, is the quantity sought in an experimental measurement.

As with the piezoelectric case, material constants are most easily determined from the initial jump in current $i(0+)$, which, from Eq. (4.16), is

$$\frac{i(0+)t_0}{A} = \left[1 - \frac{1}{\alpha}\left(1 - \frac{u}{U}\right)\right]D_0, \tag{4.17}$$

where D_0 is the initial electric displacement $E\varepsilon^+$ and α is the ratio of strained to unstrained permittivity.

Experimental studies within the elastic range have been performed on monocrystalline Al_2O_3 (sapphire) and the nonpiezoelectric z-cut of quartz. Experiments are performed with a circuit devised by Ingram [68G05] in which a low-loss coaxial cable is used for both application of the potential and monitoring the current. As shown in Fig. 4.7, at an applied potential difference of a few kilovolts, a current of about 1 mA is produced at a compression of several percent.

Constants determined from data reported in reference [68G05] and from the piezoelectric studies of X-cut quartz are shown in Table 4.3. The coefficients are found to be constant over the range of strain indicated.

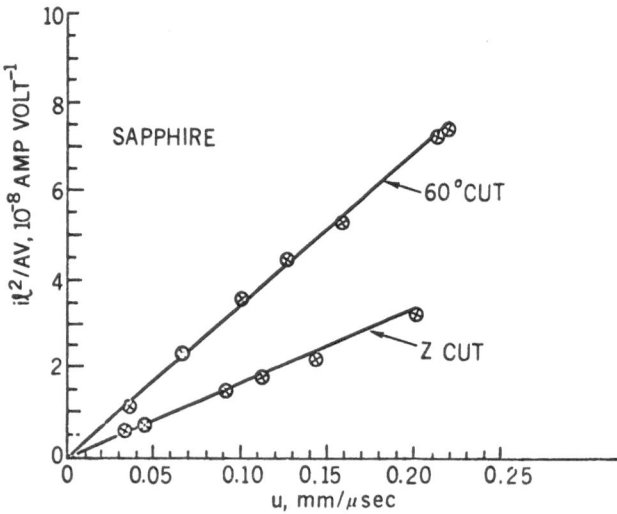

Fig. 4.7. The dielectric permittivity of impact-loaded dielectrics can be determined from current pulse measurements on disks biased with a voltage V. The magnitudes of the normalized current pulse values shown for two crystallographic orientations of sapphire are linear change with applied strain (after Graham and Ingram [68G05]).

Table 4.3. Electrostrictive constants (after Davison and Graham [79D01]).

Material/orientation	Experiments	Strain range	f_{iii} (F/m)
z-cut Al_2O_3	5	0.3–1	+58
60° cut Al_2O_3	7	0.3–2	+86
z-cut quartz	4	2.4–6.5	+6
x-cut quartz	many	···	−4

f_{iii} is the electrostrictive constant for the indicated orientation.

Similar studies can be performed above the elastic range if the hydrodynamic model is a suitable approximation to the response of the material. Such studies have provided permittivity data on polyethylene to 25 GPa [70H02], although these studies were complicated by a "shock-induced polarization" effect. In materials that exhibit shock-induced polarization, permittivities can be determined in a manner analogous to that used for piezoelectric solids [65H01, 70H02]. Hauver [70H02] has used a resonant LC circuit to determine permittivities for an organic material, *o*-nitroanisole. In shock-compression experiments on ferroelectrics, Lysne [78L05] has measured the permittivity change on a "slim loop" ferroelectric ceramic which shows a linear reversible permittivity change with stress from 0.24 to 0.88 GPa.

4.3 Shock-Induced Conduction in Elastic Dielectrics

Shock-induced conduction in piezoelectrics is differentiated from that in other dielectrics because it is observed under the unusually high electric fields produced by the piezoelectric effect in the thick-sample configuration. Shock-induced conduction observed in quartz and lithium niobate has been identified as dielectric breakdown or a prebreakdown electrical process associated with electric fields in the range of 10^7–10^8 V m^{-1} and limited defect generation. The dielectric strengths under shock loading are less than 10% of the atmospheric-pressure values. Given the high shear stress present in these experiments, it is not difficult to believe that dielectric strength could be reduced, but the physical mechanisms responsible for the observations have not been identified. In spite of interest extending over 15 yr and the conduct of a number of detailed investigations, no physical model for shock-induced dielectric breakdown has been developed. Fortunately, the breakdown for quartz and lithium niobate occurs well within the elastic range and under conditions in which the strains and electric fields can be accurately calculated.

The sample-polarity anomaly in current pulses from x-quartz shocked above the Hugoniot elastic limit gave the first indication of unusual conduction phenomena in that material [62N02]. Subsequent work [62G01, 68G04, 72G03] showed anomalies in current pulses in the minus-X orientation at all fields above a threshold stress of 1.2 GPa. The observed rapid reduction in current in an external short circuit, and intense localized luminescence indicative of localized high-current densities [62N02, 65B03, 68G04], supply strong evidence to support a breakdown phenomenon.

Certainly the most prominent feature of the breakdown process is its dependence on the polarity of the electric field relative to the shock-velocity vector. This effect is manifest in current pulse anomalies from minus-x orientation samples or positively oriented samples subjected to short-pulse loading (see Fig. 4.8). The individual effects of stress and electric field may be delineated with short-pulse loadings in which fields can be varied by utilizing stress pulses of various durations [72G03].

As shown in Fig. 4.9, these studies provide evidence that the breakdown was characterized by a fixed threshold stress of 11 GPa and a fixed threshold field of 2.8×10^7 V m^{-1}. Once the threshold stress is exceeded, the conduction is controlled by the field and is independent of the stress. The threshold field is in reasonable agreement with the field of 7×10^7 V m^{-1} below which a recovery from breakdown is observed when the field decreases due to the

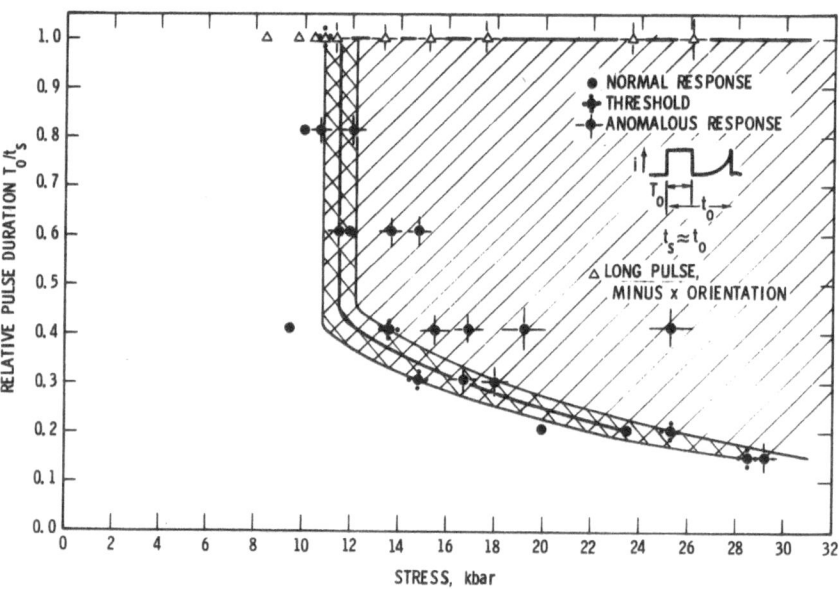

Fig. 4.8. When x-cut quartz is subjected to impact loading whose duration is less than wave transit time, an anomalous current pulse can be observed after the stress release. The diagram shows locations at which experiments were conducted and delineates the region of normal and anomalous response (after Graham and Ingram ([72G03]).

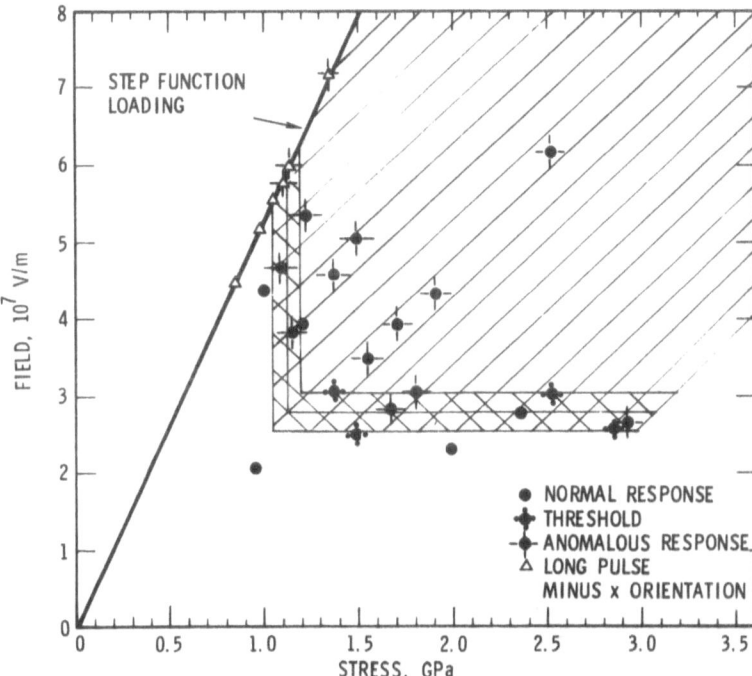

Fig. 4.9. The data of Fig. 4.8 are represented in the electric-field, stress plane to show that the anomalous response occurs above a critical stress and critical electric field. The response is found to be due to dielectric loss or shock-induced conduction (after Davison and Graham [79D01]).

internal conduction [68G04]. A time delay for breakdown has been observed that depends upon the electric field but is, again, independent of stress once the threshold stress is exceeded [75G06]. Such a delay time has also been observed in short-pulse loading experiments [75G06].

It appears that the observed breakdown must be explained in terms of the transient behavior of stress-induced defects even though the stresses are well within the nominal elastic range. In lithium niobate [77G06] and aluminum oxide [68G05] the extent of the breakdown appears to be strongly influenced by residual strains. In the vicinity of the threshold stress, dielectric relaxation associated with defects may have a significant effect on current observed in the short interval preceding breakdown.

The effect of shock-induced conduction is less distinct in ferroelectrics than in piezoelectrics but is nevertheless apparent from a number of studies. (See Davison and Graham [79D01] and Novitskii [79N03].) Differences in conduction with sample polarity, such as those seen in quartz but of opposite sign, are observed in ferroelectrics.

4.4 Semiconductors Under Large Elastic Strain

Electrical properties of semiconductors are sensitive to changes in energy band structure and impurities. Hence, it is possible to use fairly simple probes such as sample resistance, self-generated emf measurements or more sophisticated probes such as Hall voltage measurements to obtain reasonably direct information on fundamental properties. Static-high-pressure and uniaxial stress have proven to be effective in such studies since both the energy gap and relative level of critical points on the band structure can be changed with isotropic and anisotropic strain. Degeneracies in band structure can also be removed with application of anisotropic strain. Knowledge of such stress effects is essential for interpreting effects of stress on semiconductor junction devices.

Unfortunately, the sensitivity of electrical properties to lattice defects makes it unlikely that measurements above the Hugoniot elastic limit will be subject to straightforward interpretation since inelastic deformation generates copious quantities of defects of essentially unknown character. Such has proven to be the case for resistance measurements above the HEL in germanium [66G01] and silicon [72C04]. Nevertheless, the large Hugoniot elastic limits of both germanium [72G05] and silicon [71G06] permit purely elastic uniaxial strains of a few percent to be applied to samples whose electrical properties are being studied.

Kennedy and Benedick [67K02, 68K03] were successful in carrying out difficult Hall effect measurements in germanium samples explosively loaded at the upper end of the elastic range. Nevertheless, the measurements did not provide sufficient information to develop a physical interpretation.

Even the relatively simple resistance measurements under elastic shock loading cannot be confidently interpreted from shock-compression data alone; it is necessary to call upon related atmospheric and elevated pressure studies. Fortunately, a well-founded picture is available for germanium [64P01] in which theory and experiment have been well reconciled. Depending upon the strain magnitude, the electrical conductivity of germanium under [111] and [100] uniaxial strain is dominated by either anisotropic, strain-induced electron population transfer, anisotropic, strain-induced splitting of the valence band maximum, or strain-induced shifts in energy gap. Uniaxial [111] strain greatly simplifies the conduction band in that sufficiently large strains convert the multivalley conduction band to a single valley conduction band. On the other hand, the valence band becomes more complicated as the degeneracy of the maximum is lifted and energy levels are split. The properties of holes in such a strained configuration present the largest uncertainty.

Theory and Analysis

To plan and fully analyze experimental studies of the effect of adiabatic elastic strain on the electrical conductivity of germanium, C.L. Julian (for-

Table 4.4. Sources of theory and data for germanium modeling. (See Davison and Graham [79D01].)

Parameter	Author
General reference	Paige [64P01]
Population transfer model	Herring; Keyes [55H01, 60K01]
$np(0, T)$	Morin and Maita; Prince [54M01, 53P01]
M_n	Paige [64P01]
$\mu_n(S, T)$	Schetzina and McKelvey [69S02]
$M_p(S, T)$	Julian and Lane [73J03]
$\mu_p(S, T)$	Asche et al. [66A04]
Hydrostatic deformation potential	Paul, Paul and Brooks [63P01, 63P03]
b, d	Pollak and Cardona [68P02]

merly of Sandia National Laboratories) has developed a computer code, "Santa Fe," whose main subroutine, "Chili," calculates conductivity and related diagnostic parameters for [111] and [100] uniaxial strains, uniaxial stresses, and hydrostatic pressure. Sources for the theory and data incorporated in the code are summarized in Table 4.4.

As is customary, the conductivity is described by independent transport of electrons and holes such that

$$\sigma = ne\mu_n + pe\mu_p, \tag{4.18}$$

where σ is the conductivity, n and p are the numbers of electrons and holes, respectively, μ_n and μ_p are mobilities of electrons and holes, respectively, and e is the electronic charge. If the sample contains ionized impurities denoted $N_A - N_D$, where N_A is the number of acceptor ions and N_D is the number of donor ions, the np product is

$$np = n^2 - (N_A - N_D)n. \tag{4.19}$$

The np product at a strain, S_1, and temperature, Θ, $np(S_1, \Theta)$ of interest can be shown to be

$$np(S_1, \Theta)/np(0, \Theta) = R_n R_p \exp(-\Delta E_g/k\Theta), \tag{4.20}$$

where R_n and R_p are the ratios of density of states effective masses (M_n and M_p) to the $\frac{3}{2}$ power of the unstrained to strained electrons and holes, respectively, ΔE_g is the change in energy gap, and k is Boltzmann's constant.

Once values for R_n, R_p, and ΔE_g are calculated at a given strain, the np product is extracted and individual values for n and p are determined from Eq. (4.19). The conductivity can then be calculated from eq. (4.18) after the mobilities are calculated. The hole mobility is the principal uncertainty since it has only been measured at small strains. In order to fit data obtained from elastic shock-loading experiments, a hole-mobility cutoff ratio is used as a parameter along with an unknown shear deformation potential. A best fit is then determined from the data for the cutoff ratio and the deformation potential.

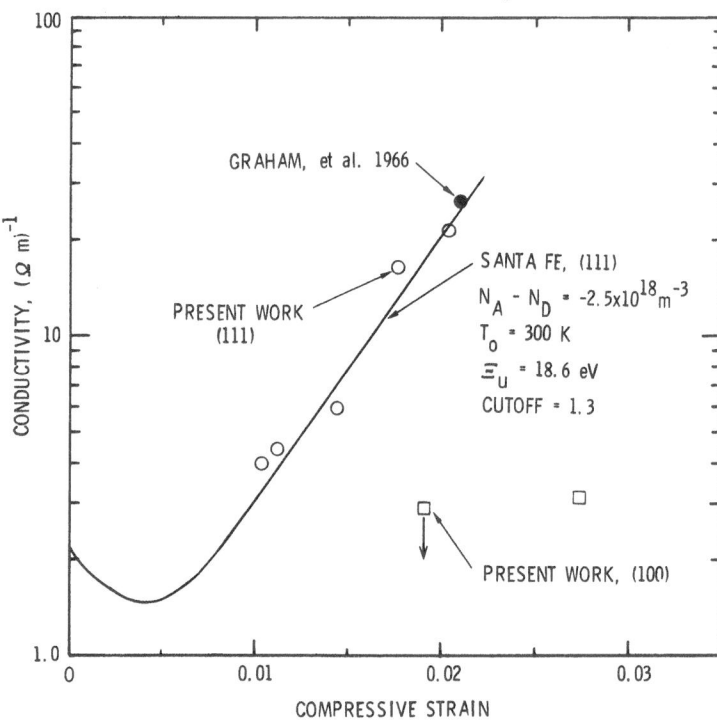

Fig. 4.10. The conductivity of uniaxially compressed (111) and (100) high purity germanium crystals leads to a determination of the shear deformation potential for the designated valley minima in the energy band (after Davison and Graham [79D01]).

The effect of [111] uniaxial strain on conductivity is illustrated in Fig. 4.10 by the calculated solid line for which the Santa Fe Code incorporated the parameters shown. (The datum points are to be described later.) At low strain the conductivity change is dominated by changes in electron mobility accompanying the electron transfer to a single (111) valley; this represents the classic piezoresistive effect for n-type germanium. Above some critical strain the electrons are completely transferred to a single (111) valley and the conductivity is dominated by the decrease in energy gap with compressive strain. Such a complete transfer of electrons to a single valley has been observed by Schetzina and McKelvey [69S02] in uniaxial stress experiments.

Experimental

There are two experimental investigations of resistivity of germanium under elastic shock compression. The work of Graham et al. [66G01] interpreted limited measurements on [111] Ge on the basis of intrinsic semiconduction without attempting to account for strain-induced changes in mobilities and effective masses. On that basis, the shear deformation potential was found to be about a factor of 2 too small. In a more recent work [79D01], Graham and Julian conducted a very similar but more detailed and carefully executed study of [111] and [100] Ge that was fully interpreted with code calculations to both plan and interpret experiments.

An extensive program to characterize conventional electrical properties of the samples at atmospheric pressure was carried out by J.D. Kennedy and R.D. Jacobson of the author's laboratory. Since each experiment required three crystals (one sample, one impactor, and one impedance-matching disk) it was possible to select sample crystals with the best-behaved electrical properties from among a large group of crystals all of which were nominally characterized by the supplier as high-purity, n-type. Preshot characterization included four-probe (Van der Pauwe) conductivity concentrations. Julian's computer code, "Sparrow," incorporated theory and experimental data from prior germanium studies [64P01] to interpret the low temperature results and arrive at the impurity concentrations. Other preshot characterizations included measurements of photoconductivity and two-probe resistivity measurements at currents from one-fourth to 1 amp. The final sample characterization is performed 500 ns prior to impact loading when the 1-amp current is pulsed on for the few microseconds of the shock loading.

The [111] orientation samples were found to be n-type with impurity carrier concentrations of from 2.4 to $8 \times 10^{18} \, m^{-3}$. The [100] samples had carrier concentrations of 14 and $18 \times 10^{18} \, m^{-3}$. Calculations with the Santa Fe Code indicated that final conductivities would not be significantly influenced by the observed range of carrier concentrations.

The data obtained from the resistance measurements are shown in Fig. 4.10. The assigned values of conductivity are limited in accuracy because the measured resistance was found to be somewhat time dependent. The [100] datum at the lowest strain was particularly so and a definite resistance value cannot be assigned to that point.

The shear deformation potential for the (111) and (100) valley minima determined by fits to the data of Fig. 4.10 are shown in Table 4.5 and compared to prior theoretical calculations and experimental observations. The deformation potential of the (111) valley has been extensively investigated and the present value compares favorably to prior work. The error assigned recognizes the uncertainty in final resistivity due to observed time dependence. The distinguishing characteristic of the present value is that it is measured at a considerably larger strain than has heretofore been possible. Unfortunately, the present data are too limited to address the question of nonlinearities in the deformation potentials [77T02].

Although the [100] data are quite limited, the shear deformation potential determined is the only measurement for this valley in germanium. At atmospheric pressure and small strains the (100) valley minimum is well above the (111) valley minima and not accessible for measurement. In the present

Table 4.5. Shear deformation potentials. (See Davison and Graham [79D01].)

Author	Theoretical/experimental	$\Xi_u(111)$ (eV)	$\Xi_u(111)$ (eV)
Goroff and Kleinman [63G01]	theoretical (Silicon)	17.3	9.6
Saravia and Brust [69S03]	theoretical	14.0	7.3
Melz [70M03]	theoretical (Silicon)	...	7.5
Graham and Julian [79D01]	experimental	19 ± 2	8.3 ± 1
Schetzina and McKelvey [69S02]	experimental	16.3 ± 0.3	...
Bulthius [68B04]	experimental	19.5	...
Balslev [66B02]	experimental	$16,2 \pm 0.4$	9.2 ± 0.3 Si
Riskaer [66R01]	experimental	18.0 ± 0.5	8.5 ± 0.5 Si
Balslev [65B04]	experimental	...	8.5 ± 0.2 Si
Dakhovskii [64D01]	experimental	16.5	
Schmidt-Tiedemann [62S01]	experimental	18.9 ± 1.7	11.3 ± 1.3 Si
Fritzsche [59F01]	experimental	19.2 ± 0.4	...

uniaxial strain experiment the (100) valley becomes the minimum point on the conduction band. The observed value agrees well with theoretical calculations on silicon.

The data indicate that elastic shock-compression resistance measurements can provide data on the effects of strain on energy gaps and deformation potentials in semiconductors. Drift mobility measurements on holes in germanium and resistivity measurements on samples with different dopings would appear to be of considerable interest.

4.5 Elastic Physical Properties

The studies of second- and higher-order elastic constants, linear and non-linear piezoelectric constants, nonlinear dielectric constants, shock-induced conduction, and the resistivity of germanium under large strain represent a broad examination of physical properties of solids under shock-compression loading. From these studies, the physical properties of solids under the least complicated conditions possible under shock loading are examined in significant detail. What general conclusions can be drawn from the results? Are the studies of physical properties significant in themselves? What conclusions can we draw concerning situations in which both elastic and inelastic shock compression is encountered?

The very large Hugoniot elastic limits in the brittle single crystals are of interest in that they apparently represent strengths close to the theoretical shear strengths. They are achievable due to the rapid loading and the hydrostatic pressure component. The measurements stand alone as significant contributions to investigations of theoretical strengths of solids.

The second- and higher-order elastic constant studies in single crystals with large Hugoniot limits have provided an examination of elastic behavior at finite strain. Ultrasonic investigations of second-order elastic constants can certainly be carried out more precisely than similar shock-compression investigations; nevertheless, the shock data are of interest. The typical shock investigation does not yield a full set of elastic constants, and, given the destructive nature of the shock experiment, it is perhaps most reasonable to rely on ultrasonic experiments for studies of such constants.

Ultrasonic investigations under hydrostatic pressure or with high-frequency surface waves can yield accurate third-order elastic constants. The distinctive aspect of the shock studies is the determination of those constants without extrapolation to large strain. The nonlinear constants are determined at large finite strain and are directly representative of the finite-strain response. The comparison between the lower strain, third-order constants and finite strain, shock data provides an evaluation of higher-order deformational features of crystal lattices. The most distinctive aspect of the shock studies are the resulting fourth-order elastic constants. These constants are largely unavailable from ultrasonic investigations, but are determined with reasonable accuracy in careful shock experiments. The fourth-order constants can provide data for more accurate theoretical descriptions of crystal lattice forces and can, in principle, be related to thermal conductivity, providing a test of lattice vibration theories.

The piezoelectric constant studies are perhaps the most unique of the shock studies in the elastic range. The various investigations on quartz and lithium niobate represent perhaps the most detailed investigation ever conducted on shock-compressed matter. The direct measurement of the piezoelectric polarization at large strain has resulted in perhaps the most precise determinations of the linear constants for quartz and lithium niobate by any technique. The direct nature of the shock measurements is in sharp contrast to the ultrasonic studies in which the piezoelectric constants are determined indirectly as changes in wavespeed for various electrical boundary conditions.

The most distinctive aspect of the shock work is the determination of higher-order piezoelectric constants. The values determined for the constants are, by far, the most accurate available for quartz and lithium niobate, again due to the direct nature of the measurements. Unfortunately it has not been possible to determine the full set of constants. Given the expense and destructive nature of the shock experiment, it is unlikely that a full set of higher-order piezoelectric constants can be determined. A less expensive investigation of higher-order constants could be conducted with the "ramp" wave or acceleration wave loading experiment described in the chapter.

The piezoelectric response investigation also provides direct evidence that significant inelastic deformation and defect generation can occur well within the elastic range as determined by the Hugoniot elastic limit. In quartz, the Hugoniot elastic limit is 6 GPa, but there is clear evidence for strong non-ideal mechanical and electrical effects between 2.5 and 6 GPa. The unusual dielectric breakdown phenomenon that occurs at 800 MPa under certain

electric field conditions provides a very dramatic indication of nonideal behavior that is initiated by mechanically induced defects.

The studies of the resistivity of shock-compressed, high purity germanium
crystals have provided the most direct atomic-level physical information on
physical processes under elastic strain of any of the works of this chapter.
Nevertheless, the shock measurements themselves are not sufficient to provide a complete physical interpretation. Other hydrostatic and physical measurements are required before the resulting deformation potentials of the
band structure could be determined. The shock studies at large strain provided another significant confirmation of the validity of the classical theory
of intrinsic semiconduction in germanium, but no new aspects of the effect at
finite strain were revealed. Perhaps the failure to identify any new nonlinear
features resulted from the dependence on other studies for the interpretation
of the results. As in the other elastic range studies, evidence for defect generation at stresses less than the elastic limit is apparent. Furthermore, the
unusual nature of the defect structures for inelastic compression causes conditions that cannot be physically interpretable.

An overall assessment of the various studies within the elastic range could
perhaps be best stated that even in these simplest shock-compression situations, the physical processes are more complex than anticipated from other
known physical studies. Unexpected responses were frequently observed.
Clearly, the elastic range as defined by the Hugoniot elastic limit does not
represent a region of purely elastic response without defect generation. To
the extent that these defects affect physical properties, such observations give
little confidence that higher-pressure experiments above the Hugoniot elastic
limit can be confidently interpreted on the basis of elementary theory.

CHAPTER 5

Physical Properties Under Elastic-Plastic Compression

In this chapter: piezoelectric crystals and polymers; ferroelectric and ferromagnetic solids; resistance of metals; shock-induced electrical polarization; electrochemistry; elastic-plastic physical properties.

Given the physical description of shock-compressed solids within their finite-strain, elastic-compression ranges presented in Chap. 4, it is possible to confidently extend experimental studies into the elastic-plastic region. In this deformation region, measured electrical responses include contributions from purely elastic nonlinear deformation, from inelastic nonlinear, rate-dependent deformations of uncertain stress tensors, and from large concentrations of defects of various complexities. Without studies in the elastic range, it is not possible to quantitatively interpret these far more complex, higher pressure studies.

As was described in Chap. 2, when stresses exceed the Hugoniot elastic limit, the typical wave profile consists of an elastic "precursor" which propagates at a speed greater than an inelastic wave which carries the materials to the peak pressure applied to the sample. For the material located between the two waves, various viscous effects can operate. For times less than the transit time of the elastic wave through the sample thickness, the sample is subject to both elastic and inelastic compression regimes whose thicknesses are time dependent. Although the nominal, heuristic description of the stress tensors within the sample describes the material as completely elastic at a fixed elastic stress level in the region between the elastic and inelastic waves, and in a fixed plastic deformation state in the region traversed by the inelastic wave, material responses are more complex. Elastic states are not as well defined as simple theory would predict. Plastic deformation is viscoplastic, introducing complex, time-dependent mechanical states.

Sample configurations in which mechanical conditions throughout the thickness are not uniform are described as "thick" samples. Sample configurations in which uniform conditions are achieved throughout the sample at a given time are described as "thin" samples. In the thin-sample configuration there are no explicit wave propagation effects on observed electrical responses. In the thick-sample configuration there are explicit wave propagation effects and material relaxations influence electrical responses.

In this chapter, the observations on piezoelectric responses of piezoelectrics within the elastic range as described in Chap. 4 are extended into ranges of shock pressures in which multiple stress waves resulting from mechanical yielding propagate within the samples. Furthermore, the behavior of a new class of piezoelectrics, the piezoelectric polymer, is described under conditions that are close to hydrodynamic. Shock compression of ferromagnetic or ferritic solids changes their magnetic states, resulting in electrical signals in external circuits. The behavior of magnetic substances that undergo pressure-induced, second-order and first-order phase transitions under shock compression are considered. Finally, perhaps the most distinctive electrical response of shock-compressed solids, shock-induced electrical polarization, is considered. Shock-induced polarization effects observed in nonpiezoelectric solids, both ionic and polymeric, are described.

5.1 Piezoelectric Responses of Crystals in the Elastic-Plastic Range

Models of piezoelectric responses of x-cut quartz in this stress range were first developed by Neilson and Benedick to interpret electrical measurements of samples subjected to high explosive loading. Their analysis was embodied in the "Neilson–Benedick Three-Zone Model" which resulted from perhaps the most perceptive materials modeling effort ever achieved in shock-compression science. The model incorporates a combination of mechanical and electrical material response features that were entirely unknown at the time and represented proposals not thought to be realistic. This model, for the first time, proposed that quartz relaxed to a state of no strength upon yielding. Such "loss" or substantial reduction of shear strength phenomena is now well accepted and was described in Chap. 2 as an elastic-isotropic stress state. In addition to loss of strength, interpretation of the responses invoked a previously unknown shock-induced electrical conduction phenomenon that is dependent on the direction of the electric field relative to the shock front.

The linear piezoelectric tensor relating polarization P_i and stress σ_j in x-cut quartz is

	σ_1	σ_2	σ_3	σ_4	σ_5	σ_6
P_1	d_{11}	$-d_{11}$	0	d_{14}	0	0
P_2	0	0	0	0	$-d_{14}$	$-2d_{11}$
P_3	0	0	0	0	0	0

In the present case only the component of polarization on an electrode normal to the x-axis, P_1, is measured. This tensor indicates that while a uniaxial strain will produce a polarization, a hydrostatic stress configuration for which $\sigma_1 = \sigma_2 = \sigma_3$ and $\sigma_4 = \sigma_5 = \sigma_6 = 0$ will produce zero polarization. Thus, measurements of piezoelectric polarization for stresses above the Hugoniot elastic limit are expected to give a direct indication of the magnitude of the shear stress.

The Three-Zone Model of Neilson and Benedick

The electrostatic configurations used to derive the piezoelectric currents are shown in Fig. 5.1. In response to a loading that exceeds the Hugoniot elastic limit, two shock waves propagate through the sample. The region between the elastic and plastic waves is considered to be elastically stressed to produce a uniform piezoelectric polarization P_H whose magnitude is directly proportional to the stress at the Hugoniot elastic limit. The material behind the plastic wave is assumed to have zero shear strength. In this configuration, the piezoelectric tensor indicates that a state of zero piezoelectric polarization is achieved behind the plastic wave. Orientations of opposite electrical polarity are assumed to exhibit radically different electrical conductivities. In the plus-x orientation, conduction occurs in region three behind the plastic wave. In contrast to this, in the minus-x orientation, region three remains an insulator while conduction occurs in elastic region two between the elastic and plastic waves. These unusual conduction effects, which are crucial for interpretation of mechanical effects, will be addressed in more detail later.

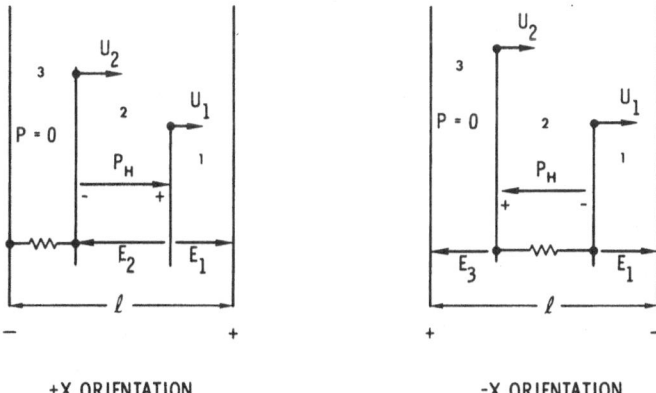

+X ORIENTATION   -X ORIENTATION

Fig. 5.1. The electrostatic configurations of the Neilson–Benedick "three-zone model" describe a piezoelectric solid subject to elastic–inelastic shock deformation which divides the crystal into three distinct zones. Zone 1, ahead of the elastic wave, is unstressed. Zone 2 is elastically stressed at the Hugoniot elastic limit. Zone 3 is isotropically pressurized to the input pressure value (after Graham [74G01]).

Plus-x Orientation

In the plus-x orientation, the region behind the plastic wave is treated as a conductor. Accordingly, in the electrical model, the left electrode is moving with the velocity of the plastic wave. Otherwise, the analysis proceeds as in the case of the elastic-dielectric. For convenience it is assumed that $\varepsilon_3 = \varepsilon_2 = \varepsilon_1$. The thicknesses of the two dielectric regions are $l_1 = l$ and $l_2 - (U_1 - U_2)t$. Solution for the current is then

$$il/A = \frac{(U_1 - U_2)P_H}{(1 - U_2 t/1)^2}, \quad 0 < t < t_0. \tag{5.1}$$

The current pulse described in Eq. (5.1) is markedly different from that describing the elastic-dielectric. Particular features of the solution may be characterized by solutions for current i_1 at times $t = 0$ and i_f at time t/U_1. They are

$$i_i l/A = (U_1 - U_2)P_H, \tag{5.2}$$

and

$$i_f l/A = \frac{(U_1 - U_2)P_H}{(1 - U_2/U_1)^2}. \tag{5.3}$$

Equation 5.2 indicates that i_i is much lower than the corresponding value for the elastic-dielectric, while Eq. (5.3) indicates that i_f is significantly greater than the corresponding value for the elastic-dielectric. Since both of these current values depend upon the velocity of the plastic wave, measurements of these currents can, in principle, be used to evaluate the velocity of the plastic wave. Observation of a waveform described by Eq. (5.1) would confirm the presence of a second wave behind which the material is conductive.

Minus-x Orientation

Analysis of the minus-x orientation must follow a somewhat different procedure since the electrically conducting elastic region between the two waves separates two insulating regions in contact with the electrodes. The typical local history of a piezoelectric "unit cell" in that region includes the following events: (1) initial application of elastic strain produces a piezoelectric polarization on the unit cell; (2) local conductivity provides free charge, which in turn becomes bound charge of equal magnitude and opposite polarity to the original piezoelectric polarization; and (3) the release of piezoelectric polarization in the insulating state leaves uncompensated bound charge on the unit cell which is a polarization state of opposite sign to the initial piezoelectric polarization.

Based on the preceding local history, the boundaries at the elastic and plastic wave fronts are characterized by bound surface charges of equal magnitude and opposite sign to the initial piezoelectric states. The polarization in the region behind the plastic wave P_3, has magnitude equal to the change in piezoelectric polarization and a sign corresponding to the change in piezoelectric polarization. Thus, it follows that $D_1 = \varepsilon E_1$, $D_2 - 0$, and $D_3 = \varepsilon E_3 + P_3$. Two equal and opposite discontinuities in displacement at the elastic and plastic wave fronts lead to the condition that $D_1 = D_3$. From this point the analysis proceeds as in the previous models. The left electrode moves with particle velocity u, and the thicknesses of the two dielectric regions are $l_1 - l - U_1 t$, and $l_3 = (U_2 - u)t$. The solution for current is then

$$il/A = \frac{-(U_2 - u)P_3}{[1 - (U_1 - U_2 + u)t/1]^2}, \quad 0 < t < t_0. \tag{5.4}$$

Solutions for i_i and i_f are then

$$i_i l/A = -(U_2 - u)P_3, \tag{5.5}$$

and

$$i_f l/A = -(U_2 - u)\left(\frac{U_1}{U_2 - u}\right)^2 P_3. \tag{5.6}$$

With the present sign convention, a predetermined minus-x orientation polarity leads to a negative current in the elastic-dielectric model. Since P_3 corresponds to the change in polarization from the initial state P_H, the sign of the current indicated in Eqs. (5.4)–(5.6) required that, if $P_3 < P_H$, the current is positive. Accordingly, observation of a nonzero current from minus-x orientation samples provides direct evidence for an insulating region behind the plastic wave; the sign and magnitude of the current provide direct evidence for the sign and magnitude of the change in polarization in the region behind the plastic wave.

Experimental Observations: x-Cut Quartz

The experimental study of x-cut quartz and lithium niobate in elastic-inelastic stress regimes was carried out with the impact-loading technique described for elastic crystals as described in Chap. 4. Very high pressure loading was also achieved with high explosive, plane-wave loading systems [62N01]. Representative current-versus-time records observed over a stress range from 2.5 to 9 GPa are shown in Fig. 5.2. The two end members shown in the figure as (a) and (f) represent the extreme differences in response. The lower stress response is that typical of the elastic-dielectric response, while the higher stress response is that typical of the three-zone model. The intermediate stress responses represent distortions from the elastic dielectric response. In the stress region below about 6 GPa, no evidence has been found for a multiple wave structure characteristic of mechanical yielding. The electrical responses indicate, however, that stress relaxation and shock-induced conduction are the likely causes for the observed waveforms.

In Fig. 5.2(f) (the highest stress) an early time current "spike" is apparent. Other records show this behavior more clearly. The impact-time spike shows

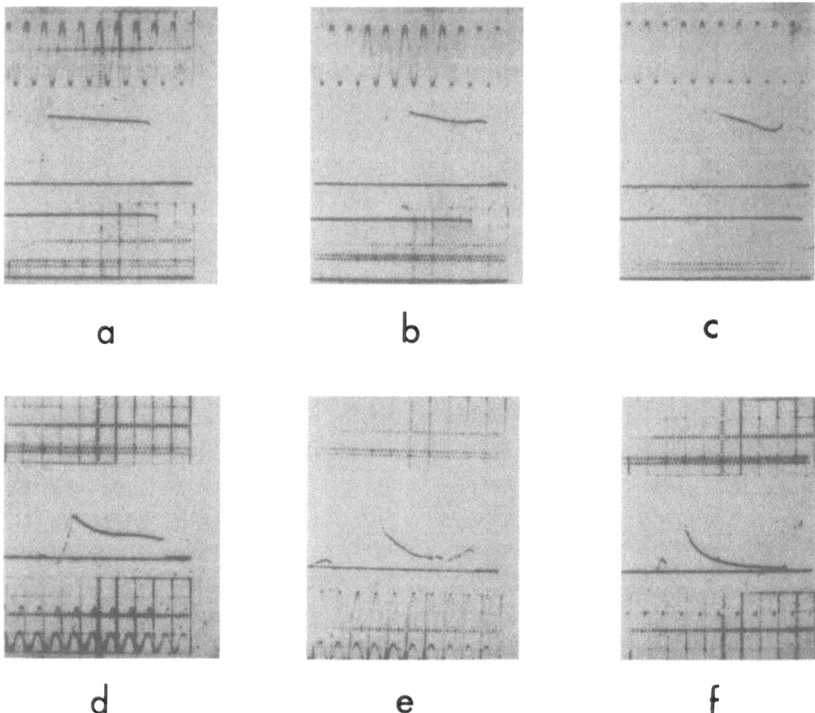

Fig. 5.2. Current-versus-time records for x-cut quartz impact loaded to stresses of 2.5, 3.9, 4.5, 5.9, 6.5, and 9.0 GPa are shown, illustrating the drastic changes occurring with mechanical yielding and conduction. Time increases from right to left. The current pulses are in the center of each record and are characterized by a brief horizontal trace (zero current before impact) followed by a rapid jump to a current value (after Graham [74G01]).

a peak in current that decays to an equilibrium value in about 10 ns. Such a relaxation is not predicted by the three-zone model and is thought to represent a delay time required to initiate dielectric breakdown in the crystal. It should be recognized that until the piezoelectric polarization is established, electric fields are not present. Thus, an initial transient current is to be anticipated.

A limited number of minus-x orientation samples were impact loaded in the vicinity of the Hugoniot elastic limit at stresses from 5.9 to 6.7 GPa. The principal observation of these experiments was that positive currents were observed from negative polarity disks when a stress of 5.9 GPa was exceeded. Such an observation confirms that quartz responds as predicted by the model, and that the elastic limit is in the vicinity of 6 GPa.

Interpretation of the experimental study of quartz leads to the conclusions that below 6 GPa and greater than 2.5 GPa x-cut quartz responds as an approximation to the elastic-dielectric model, but that there are very significant deviations from that simple model. Above 6 GPa, quartz responds as an approximation to the three-zone model, but there are significant deviations from that simple model. The agreement of the general features of the observed current pulses with the three-zone model provides direct evidence for the low piezoelectric polarization state in the region above the elastic limit. Based on the observations, the stress-versus-volume relation for shocked quartz has been established and it shows that x-cut quartz retains a small, but detectable strength at high pressure. As the conclusion is based solely on electrical response and not mechanical wave responses, it provides a remarkable confirmation of the unusual nature of mechanical yielding in this material.

Experimental Observations: Lithium Niobate

Observations of current pulses from shock-loaded, x-cut quartz in the vicinity of and above the Hugoniot elastic limit provided rather remarkable confirmation of the nature of the phenomena resulting from mechanical yielding and shock-induced conduction. Lithium niobate provides another opportunity to test the generality of the models.

High pressure explosive loading was carried out on both z- and y-cut crystals at pressures between about 25 and 60 GPa ([83S01, 77S01]). The z-cut crystals responded in the plus-x orientation with current pulse wave shapes as predicted by the three-zone model. Nevertheless, limited experiments in the minus-z orientation of lithium niobate do not show the positive currents expected from the three-zone model.

The y-cut crystals showed little, if any, output signal under the same conditions for which the z-cut crystals were studied. In this case it should be observed that the y-cut crystals exhibit higher elastic limits and much higher piezoelectric polarizations than the z-cut crystals. These conditions result in much higher electric fields in the elastic region, and these fields are apparently sufficiently large that the crystals were completely conductive internally in the region between the elastic and plastic waves.

It is interesting that z-cut lithium niobate has a two-wave structure to pressures approaching 100 GPa. This feature makes the material an attractive detector for the arrival time of high pressure shock pulses [90B02]. The elastic-plastic region was found to behave as expected from a heterogeneous yielding model of intense, hot local deformation as described by Grady [77G07]. Syono has shown that a high pressure phase transformation is encountered at pressures of 33 GPa [88S03].

5.2 Piezoelectric Polymers

The science and technology of piezoelectric materials has long been dominated by the availability of specific materials with particular properties. Piezoelectric polymers are the most recent class of piezoelectrics developed, and their characteristics are significantly different from the more conventional piezoelectrics. A careful study of the piezoelectric properties of one of these materials, polyvinylidene difluoride (PVDF), has been carried out under high pressure shock compression.

The piezoelectric effect was first identified by Jacques and Pierre Curie in France in 1868 in studies on tourmaline, a naturally occurring crystal that is both piezoelectric and pyroelectric [64C01]. In modern times, naturally occurring crystalline α-quartz became the most widely used piezoelectric material based on availability and precise, reproducible properties. During World War II, technology was developed to grow large crystals synthetically to support the widespread use of quartz in electronic communication devices. Although widely used, the low efficiency of quartz for conversion of electrical to mechanical energy (its electromechanical coupling coefficient) led to the development of ceramic ferroelectrics based on lead-zirconate-titanate of various compositions with substantially larger efficiencies. These polycrystalline ceramics typically do not have highly reproducible responses and are too lossy for precise applications. Other ferroelectrics, such as lithium niobate and lithium tantalate, have become commercially available in single crystal form in the 1970s and are widely employed in various devices, including shock-compression gauges.

The most common piezoelectric polymers are PVDF, based on the monomer $-CH_2-CF_2$ and copolymers of PVDF with C_2F_3H. Although there are many sources for materials that are nominally piezoelectric, sources for reproducible materials are limited.

In 1969, Kawai [69K01] first reported that semicrystalline PVDF film became piezoelectric after mechanical stretching and the subsequent application of an electric field. Kepler [78K01] identified the three stable polymorphs of crystalline PVDF. Recognizing the need for a reproducible material for the destructive conditions of shock compression, Bauer [82B01] developed technology to process PVDF such that its physical properties exhibit reproducibility approaching that of piezolectric single crystals. Based on the Bauer processing methods, it is now possible to study the properties of PVDF under high pressure shock compression. The material has the potential for a new shock-wave profile sensor with revolutionary characteristics.

PVDF Film Material

In order to anticipate problems and to interpret observations under the extreme conditions of shock compression, it is necessary to consider structural and electronic characteristics of PVDF. Although the phenomenological piezoelectric properties of PVDF are similar to those of the piezoelectric crystals, the structure of the materials is far more complex due to its ferroelectric nature and a heterogeneous mixture of crystalline and amorphous phases which are strongly dependent on mechanical and electrical history.

Prior to the mechanical stretching into a film, PVDF resin shows only limited crystallinity. Upon stretching, a concentration of the polar β-phase is developed. As it is the mechanical deformation that induces the electrically active polar phase, the electronic properties are controlled to first order by the extent and uniformity of the stretching. For this reason, biaxially stretched PVDF has a significantly higher polar phase concentration than uniaxially stretched film. It should be observed that instabilities encountered in stretching will lead to local regions in the film where properties vary from those in other locations. High quality, biaxially stretched PVDF is thought to contain about 50% crystalline material.

It is known that mechanical and physical properties of the amorphous and crystalline phases differ significantly [80T01]. For this reason, it is antici-pated that properties of the mechanically and electrically treated film will depend explicitly on its history. Shock-compression measurements such as those carried out on amorphous materials in a thick form [80M01] will not prove characteristic of thin, treated films.

Although high concentrations of β phase are produced by stretching, the crystallites are randomly oriented. As in typical ferroelectrics, the crystallites are aligned in a direction normal to the film surface by an externally applied electric field. Polymers are well known to develop space charge on surfaces under the presence of such a field, and in the present situation the inhomo-geneous distribution of crystalline material can cause local sites for space charge within the bulk of the film. The space charge problem can induce electrical effects typical of electrets, which can be quite significant. For this reason, special electrical poling procedures must be employed to achieve a uniform, precisely specified remanent polarization. The value of this polariza-tion controls the piezoelectric properties of the polymer.

The Bauer poling process [86B01, 87B01] subjects the film to a cyclically varying electric field which is slowly increased in magnitude through many positive and negative polarity cycles. This electrical treatment effectively re-moves space charge from the material. A reproducible remanent polarization of 9 $\mu C\,cm^{-2}$, which is uniform throughout the thickness of the film, can be readily achieved with the Bauer process. The electrical polarization versus applied electric field of a high quality film prepared with the Bauer method is shown in Fig. 5.3. For the shock-compression studies, each PVDF sample is furnished with such an individual description, which provides a direct mea-sure of the electrical and mechanical state of the sample.

Copolymer compositions differ from PVDF in that the crystalline β phase is obtained without mechanical deformation. Thus, various thicknesses of the material can be readily produced. Unfortunately, a reproducible copolymer is not yet commercially available.

In this work, the response of PVDF is studied in a cooperative effort with Francois Bauer of the Institut Saint Louis, France, L.M. Lee of the Ktech Corporation of Albuquerque, New Mexico, and M.U. Anderson, R.P. Reed, and L.M. Moore of Sandia National Laboratories. The basis of the work is

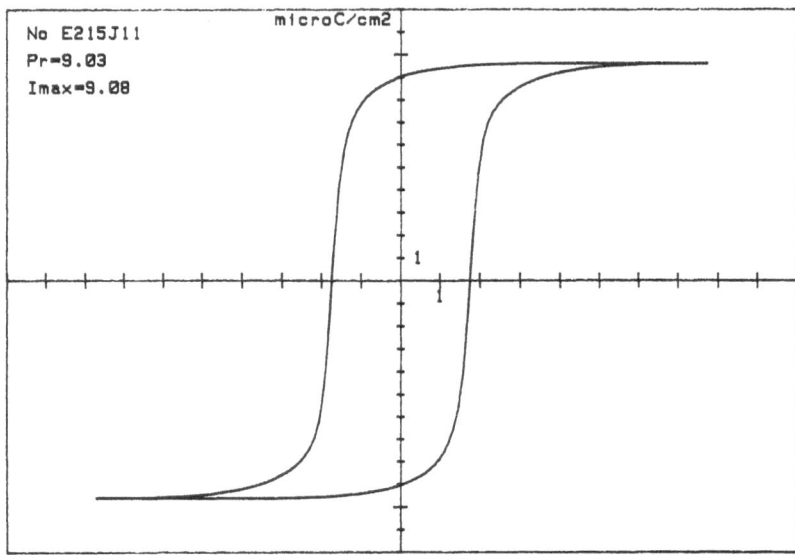

Fig. 5.3. High quality, biaxially stretched PVDF ferroelectric film can be electrically poled to a remanent polarization of over 9 μC cm^{-2} with the Bauer cyclic poling technique. The polarization versus electric field of a typical sample is shown.

a standardized, reproducible PVDF material produced commercially by the Metravib RDS Company, France, under license of the Institut Saint Louis.

Thin and Thick Sample Configurations

In Chap. 4 piezoelectric crystals were studied with thicknesses such that the few nanosecond time resolution of the current pulse measurement was short compared to the time for stress waves to propagate through the thickness of the crystal. This mechanical and electrical arrangement is described as a thick-sample, current-mode configuration. For PVDF the studies will be carried out in what is described as a thin-sample, current-mode configuration. The thickness of the standardized PVDF film used in the studies is nominally 25 μm with wave transit times that vary from 5 to 10 ns, depending upon the stress amplitude. The characteristics of the thick- and thin-current-mode arrangements as well as charge-mode arrangements are summarized in Fig. 5.4.

For times less than the transit time of the wave, the current is proportional to the stress at the input electrode in a linear approximation. For times greater than the wave transit time, the current is proportional to the stress difference between the electrodes. Thus, the thin-film nature of PVDF provides a means to measure stress differences, and, given mechanical tolerances that limit loading times to a few nanoseconds, measurements are difficult to

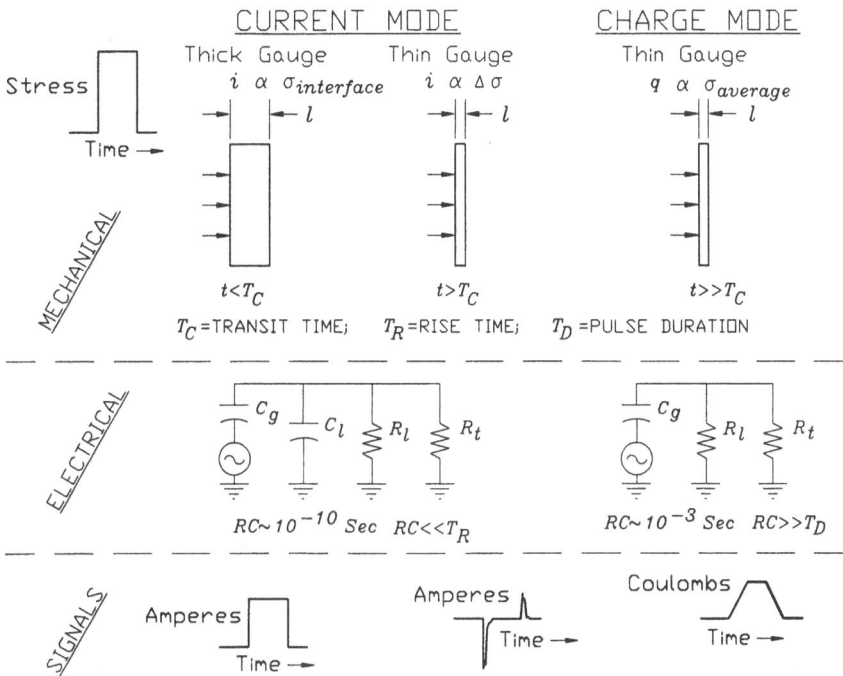

Fig. 5.4. The electrical signals from shock-compressed piezoelectric solids depend explicitly on the electrical circuit and mechanical arrangement (the sample thicknesses). In the current mode (low electrical impedance), the current pulse either follows the loading as a close analog, or, in the thin mode of PVDF, follows the derivative of the stress pulse in time.

accomplish during wave transit time. For times significantly greater than wave transit time, the current provides a direct measure of stress rate.

Experimental Arrangement

The experimental arrangement is a modification of that used for earlier studies of piezoelectric crystals. In this case the thin-film nature of the sample presents special problems in the loading. The arrangement is shown in Fig. 5.5. Standard target and impactor configurations are used with the single crystals z-cut quartz or z-cut sapphire. A highly reproducible impact stress can be obtained at stresses less than their Hugoniot elastic limits of about 12 and 16 GPa, respectively. In this configuration, the peak stress is achieved by wave reverberations between the target and impactor, and the peak stress is not dependent on the shock-compression properties of the PVDF. For stress amplitudes less than about 1 GPa, a lower shock impedance material such as the polymer Kel F is employed. Its impedance is close to that of PVDF [80M01]. At higher pressures, tungsten carbide can be employed for the

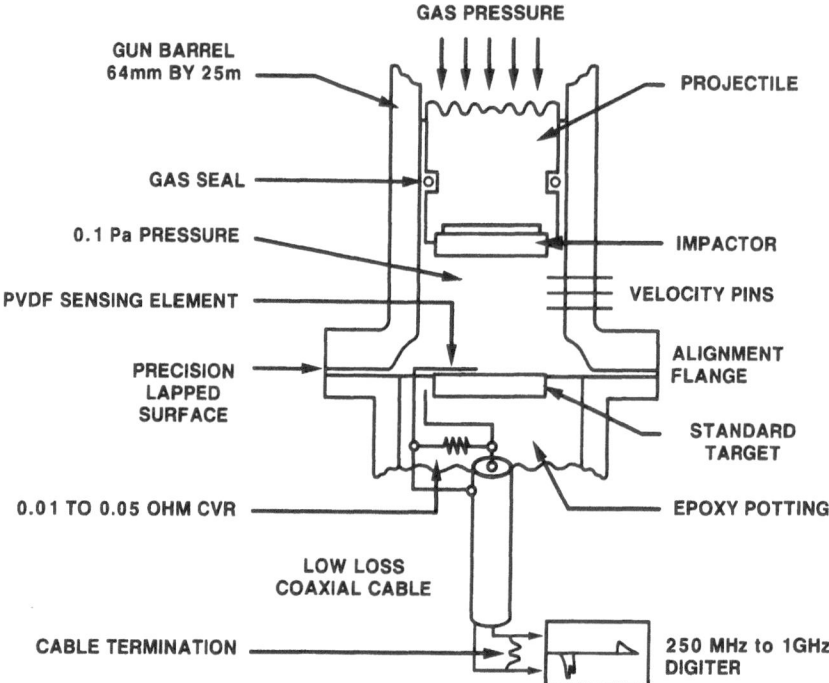

Fig. 5.5. The electrical response of piezoelectric polymers under shock loading is studied experimentally by placing the thin PVDF element on the impact surface of a standard target, either the polymer, Kel F, z-cut quartz, or z-cut sapphire. The impactor is typically of the same material. The current pulse is recorded on transient digitizers with frequency responses from 250 to 1000 MHz.

impact loading. In this case the shock-compression properties are not sufficiently well known to serve as a precise standard. At intermediate pressures, copper can be used as a well known standard.

To study the lowest pressure region from about 10 MPa, the acceleration loading pulse method previously used by Setchell [88S01] has been employed. In this case the slowly rising stress pulse from high quality fused quartz is in the form of a "ramp" in pressure. Hence, a continuous response can be determined to stresses up to about 3 GPa.

To examine the response of PVDF to a higher pressure, ramp-type loading, an experimental arrangement with a ceramic "Pyroceram" similar to that of the fused quartz loading was employed. In this case, the loading wave transmitted through the Pyroceram is only approximately known and a simultaneous measurement must be carried out of the wave transmitted through the pyroceram and the response of the PVDF.

For the shock loading of the present experiment the currents achieved are typically tens of amperes. To keep the voltages sufficiently low to measure

conveniently, low inductance, current-viewing resistors with resistances between 0.01 and 0.5 Ω are used for the electrical load. Current pulses are recorded on either 350 MHz digitizers (LeCroy 6880E, 742 ps digitizing rate)
or, more recently, 1 GHz digitizers (Tektronix DSA 602, 500 ps digitizing
rate).

Typical current-versus-time responses recorded with the various impactor-
target materials are shown in Fig. 5.6. In each case the shape of the pulse

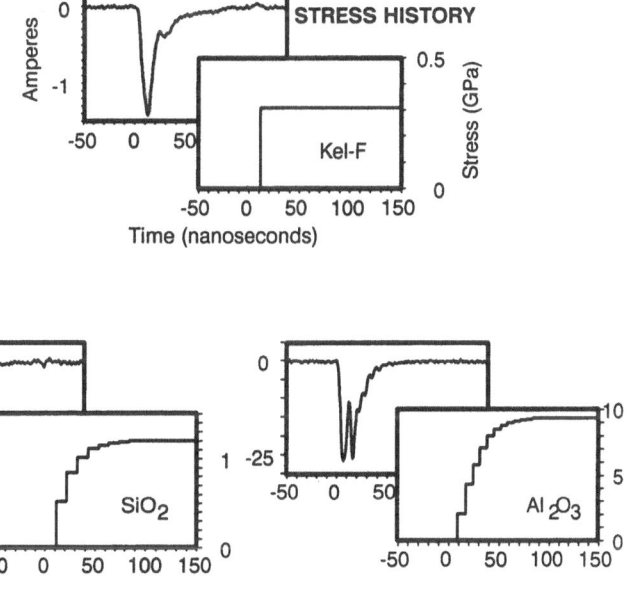

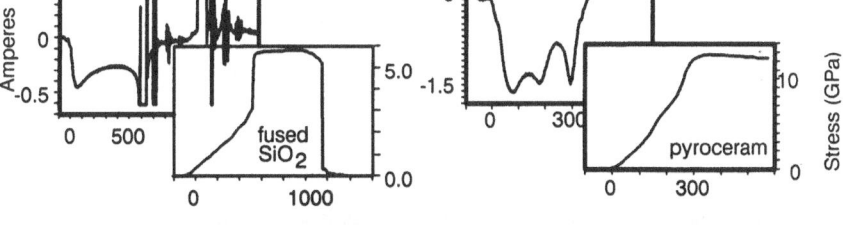

Fig. 5.6. Typical current-time responses from impact-loaded PVDF are shown for
samples on the standard materials indicated. In each record the upper traces are the
full record showing short duration negative and positive current pulses due to loading
and release in the standard. Time increases from left to right. The detail of each pulse
depends upon the shock impedance of the materials. In each record, enlarged views of
loading and release pulses are shown.

differs due to the various shock impedances of the standards. In the case of Kel F, the current upon loading is characterized as a single pulse with only a small wave reverberation. By using different materials to achieve the same peak pressure, the response of PVDF under different loading paths is studied.

As the current pulse is largely dominated by the stress differences, a short duration current pulse is observed upon loading with a quiescent period during the time at constant stress. With release of pressure upon arrival of the unloading wave from the stress-free surface behind the impactor, a current pulse of opposite polarity is observed. The amplitude of the release wave current pulse provides a sensitive measure of the elastic nonlinearity of the target material at the peak pressure in question.

Results and Discussion

The piezoelectric charge associated with a particular peak pressure is obtained upon integration of the current pulse. The charge data are shown in Fig. 5.7 to peak pressures to 9.4 GPa. The data represent loadings over the various paths. The data on peak charge observed for the various loading paths confirm that the charge is piezoelectric in character. There is no evidence for nonlinear ferroelectric behavior. The agreement between shock

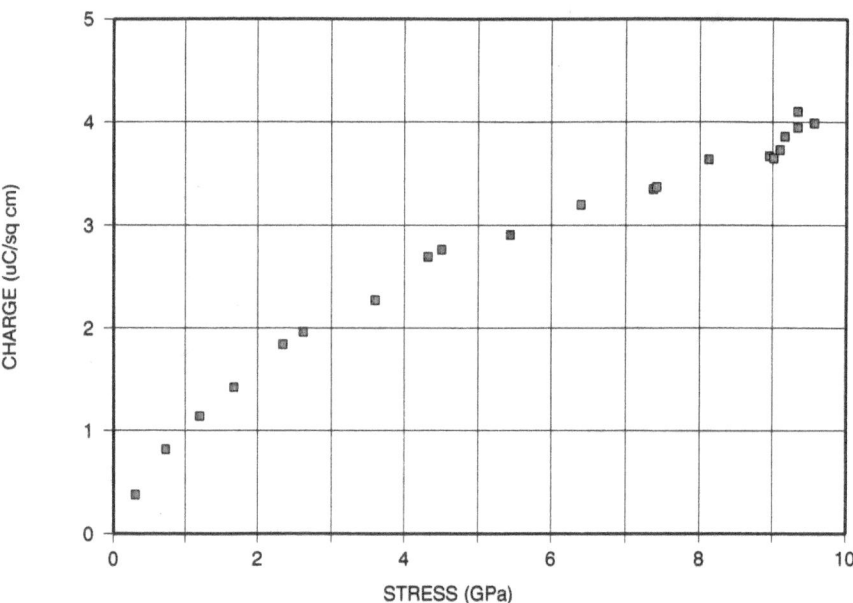

Fig. 5.7. The electrical charge observed at various peak stresses is shown for PVDF. The indicated behavior is nonlinear, but reproducible and independent of loading path.

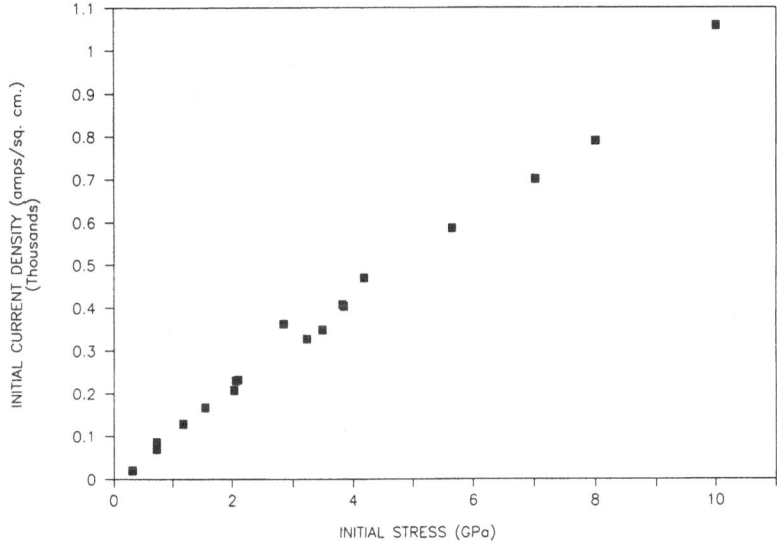

Fig. 5.8. If a peak current value is recorded before wave transit time in the PVDF, the current is a measure of response in the thick, current mode. The observed currents are found to be approximately linear with stress.

loading and acceleration pulse loading from fused quartz is particularly significant.

For those cases in which a peak current value is achieved prior to wave transit time in the samples, the current provides a direct measure of the piezoelectric polarization in a time of a few nanoseconds. The data from such experiments are shown in Fig. 5.8. A relation showing a remarkable linearity with shock pressure is shown.

Characteristic responses are readily obtained at pressures higher than 10 GPa, but differences have been observed with different loading arrangements. Piezoelectric responses at higher pressures are currently under study [92B01]. Dielectric relaxation and shock-induced conductivity may be involved.

The compressibility of polymers is strongly nonlinear at pressures of a few GPa. In order to consider the nonlinearity of the piezoelectric effect at shock pressure, it is of interest to consider the piezoelectric polarization in terms of the volume compression as shown in Fig. 5.9. The pressure-versus-volume relation for PVDF is not accurately known, but the available data certainly provide a relative measure of changes in compressibility. When considered versus volume, the piezoelectric polarization is found to to be remarkably linear. Thus, large volume compression does not appear to introduce large nonlinearities. Such a behavior will need to be considered when the theory of piezoelectricity for the heterogeneous piezoelectric polymer is developed.

As shown in Chap. 4, the piezoelectric current is directly related to the

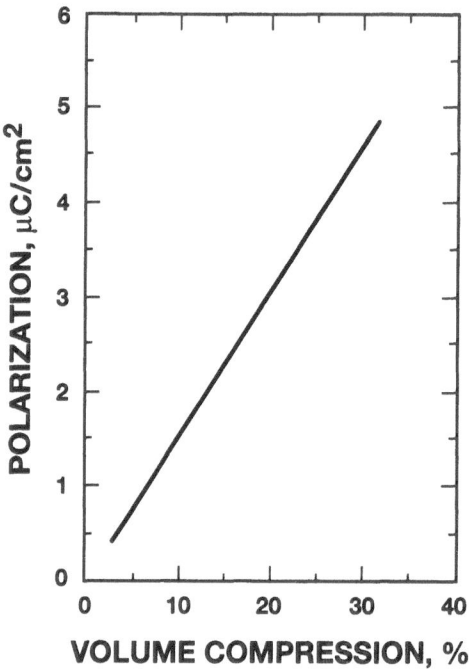

Fig. 5.9. The piezoelectric polarization of Fig. 5.7 is found to increase with volume compression. Hence, the large decrease in change of charge with stress is a manifestation of the highly nonlinear stress-volume relation of PVDF, not nonlinear piezoelectricity.

wavespeed through the relationship:

$$i_i = \alpha(1 - u/U)P_\sigma A/t_0, \tag{5.7}$$

where i_i is the initial current, α is the ratio of stressed to unstressed permittivity, P_σ is the polarization, A is the area, and t_0 is the wave transit time determined by the wavespeed and thickness.

Although the effects of dielectric constant change and strain have a strong effect on the current during wave transit time, the current at a time about $\frac{1}{2}$ transit time is close to the value for the linear relation. Thus, based on Eq. 5.7, the wavespeed can be computed from the measured current and the measured polarization data. The approximate agreement between currents calculated from the polarization data and the wavespeed data confirms that the wavespeed values currently available are reasonable.

The softening point of PVDF at atmospheric pressure is about 160 °C, but Samara has shown this temperature to increase substantially with pressure [88S02, 89S02]. It appears that under the high pressure of the shock loading the softening point is higher than the temperatures achieved due to the adiabatic compression. Upon unloading, however, the release would be expected to show evidence for softening above some critical pressure.

The piezoelectric response of PVDF at pressures greater than 10 GPa is still under study but provides an interesting expansion of studies on single crystals and promises to provide new insight into high pressure physical processes in polymers. An extension of existing work on PVDF into copolymer compositions would prove very helpful as a fully crystalline polymeric state is achieved. Furthermore, in the copolymers, the ferroelectric Curie point is at a temperature lower than the softening point. The opposite is true for PVDF. Studies on copolymers must await the development of a reproducible source of material [90B03].

5.3 Ferroelectric Solids

In this book those ferroelectric solids that respond to shock compression in a purely piezoelectric mode such as lithium niobate and PVDF are considered piezoelectrics. As was the case for piezoelectrics, the pioneering work in this area was carried out by Neilson [57A01]. Unlike piezoelectrics, our knowledge of the response of ferroelectric solids to shock compression is in sharp contrast to that of piezoelectric solids. The electrical properties of several piezoelectric crystals are known in quantitative detail within the elastic range and semiquantitatively in the high stress range. The electrical responses of ferroelectrics are poorly characterized under shock compression and it is difficult to determine properties as such. It is not certain that the relative contributions of dominant physical phenomena have been correctly identified, and detailed, quantitative materials descriptions are not available.

The contrast in knowledge is a result of the degree of complexity of materials properties: elastic piezoelectric solids have perhaps the least complex behaviors, whereas ferroelectric solids have perhaps the most complex mechanical and electrical behaviors of any solid under shock compression. This complexity is further compounded by the strong coupling between electrical and mechanical states. Unfortunately, much of the work studying ferroelectrics appears to have underestimated the difficulty, and it has not been possible to carry out careful, long range, systematic efforts required to develop an improved picture.

From the mechanical viewpoint, ferroelectrics exhibit unsteady, evolving waves at low stresses. Waves typical of well defined mechanical yielding are not observed. Such behavior is sensitive to the electrical boundary conditions, indicating that electromechanical coupling has a strong influence. Without representative mechanical behavior, it is not possible to quantitatively define the stress and volume compression states exciting a particular electrical response.

From the electrical viewpoint, stress-induced changes in remanent polarization are known to exist. The underlying changes in domain alignment are irreversible. Accordingly, electrical properties will be both stress and electric field dependent as stress waves traverse a given sample material. At sufficiently high stresses, electrical conduction is known to exist. The effect is initiated by a combination of stress and the large magnitude of electrical fields.

Given the complications of strongly nonlinear mechanical and electrical behaviors in a strongly coupled mode along with electrical conduction effects, it is not difficult to appreciate why the physical processes are poorly understood.

Work in this area is succinctly summarized by Davison and Graham [79D01], which should be consulted for details. Early work by Reynolds and Seay [61R01], Doran [68D02], and Halpin [66H02] provided the first detailed data on responses. The more complete work on the response of a wide variety of ferroelectrics has been carried out by Bauer [76B02, 82B01]. Novitskii [77N02] has also reported extensive studies of a number of ferroelectrics.

5.4 Ferromagnetic Solids

Magnetic properties of solids can be significantly altered by high pressure shock compression, from both shear stress and hydrostatic components. Those ferromagnetics whose Curie temperatures are strongly dependent on hydrostatic pressure or volume compression can be forced through second-order phase transitions with detectable changes in volume compressibility. These ferromagnets are typically iron alloys that are stable in the fcc phase. Those ferromagnetics that are stable in the bcc phase typically experience structural transformations to denser phases with reduced magnetization. Ferromagnetics and ferrimagnets exhibit stress-induced magnetic anisotropy coefficients that can alter the direction of remanent magnetization with the application of shear stress [51B01]. The first two magnetic effects are studied in this section under the conditions of high pressure shock compression.

Second-Order Curie Point Transitions

Unusual physical properties that are strongly correlated with unusual magnetic properties are characteristic of the alloys of about 30%–40% Ni in Fe. It has been known for many years that the Curie temperature and saturation magnetization of these alloys show an enormous sensitivity to pressure, reflecting a strong volumetric dependence of the magnetic interactions [63K04]. This strong pressure dependence has allowed a number of investigations of the volumetric dependence of the magnetic order with the use of relatively low (500 MPa) pressures. From these previous investigations it appears, as we will show, that the change in compressibility associated with the change in magnetic interactions is large enough to be readily measured in shock-wave compression experiments. It is the object of the present section to report an investigation of the shock-wave compression of a 30% Ni–70% Fe alloy in the fcc phase that has resulted in the identification and determination of the thermodynamic properties of the pressure-induced ferromagnetic to paramagnetic transition. The work represents the first identification of a pressure-induced, second-order phase transition under shock-compression loading [67G01].

The transition from a ferromagnetic to a paramagnetic state is normally considered to be a classic second-order phase transition; that is, there are no discontinuous changes in volume V or entropy S, but there are discontinuous changes in the volumetric thermal expansion β, compressibility k, and specific heat C_p. The relation among the variables changing at the transition is given by the Ehrenfest relations,

$$\Delta k_T = \Delta \beta \left(\frac{d\theta}{dP} \right), \tag{5.8}$$

and

$$\Delta C_p = TV\Delta \beta \left(\frac{d\theta}{dP} \right)^{-1}, \tag{5.9}$$

where Δ indicates the change occurring at the transition, k_T is the isothermal compressibility $-V^{-1}(dV/dP)_T$, and $d\theta/dP$ is the coefficient of Curie temperature change with pressure. It is clear from Eqs. (5.8) and (5.9) that a measurement of Δk_T and $d\theta/dP$ will result in a complete description of the thermodynamic properties of the transition.

The alloys of from 30% to 40% nickel in iron are noted for their unusual volumetric behavior. For example, it is well known that the thermal expansion of these alloys is anomalously low, with the Invar composition (36-wt% Ni) having a thermal expansion close to zero at room temperature. Furthermore, the atmospheric pressure compressibilities are anomalously large, whereas the atomic lattice spacing and density data show strong departures from Vegard's law in this same composition range.

For alloys containing up to about 27% Ni in Fe, the equilibrium phase at room temperature is bcc. However, in the neighborhood of 30% Ni either the fcc or bcc phases can be obtained at room temperature as the result of various heat treatments. For nickel concentrations greater than 30%, the structure is fcc. Hence, the unusually large volumetric phenomena are characteristic of the fcc phase.

The pressure sensitivity of the magnetic properties of the Invar alloys is indicated by extensive measurements of the coefficient of saturation magnetization change with pressure $M_s^{-1}dM_s/dP$ for various compositions as shown in Fig. 5.10. The exceedingly large values in the 30%–40% Ni range are evident and much in excess of the values for iron and nickel. The 30-wt% Ni composition in the fcc phase is the most sensitive to pressure, whereas this

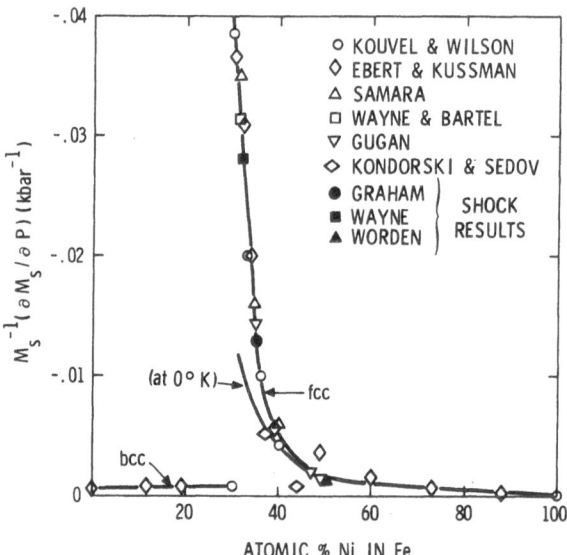

Fig. 5.10. The pressure dependence of saturation magnetization for iron-nickel alloys shows a strong pressure dependence in the neighborhood of the "Invar" alloys (28.5 to 40-at. % nickel in the fcc phase). The shock data shown are in excellent agreement with the static high pressure data (after Wayne [69W01]).

composition in the bcc phase is insensitive to pressure although strongly ferromagnetic.

Curran [61C01] has pointed out that under certain unusual conditions the second-order phase transition might cause a cusp in the stress-volume relation resulting in a multiple wave structure, as is the case for a first-order transition. His shock-wave compression measurements on Invar (36-wt% Ni-Fe) showed large compressibilities in the low stress region but no distinct transition.

Shock-wave loading is accomplished by the planar impact technique as shown in Fig. 5.11. In this investigation, both symmetrical impact and quartz gauge "front–back" configurations were used. Impacting with a quartz gauge enables a measurement of the stress and particle velocity imparted to the sample at the impact face. These quartz gauge data provide independent measurements of the shock-wave behavior from that obtained at the conventional measuring station at the back side of the specimen and provide a wave transit time measurement not influenced by nonplanarity. The redundancy achieved allows cross comparisons of data from a single experiment and greatly increases the confidence in the results since it serves as a quantitative measure of errors in the entire data recording and data reduction process as well as a verification of the assumptions made in reducing the data.

Commercially available 28.5-at. % Ni–Fe and 35-at. % Ni–Fe alloys were

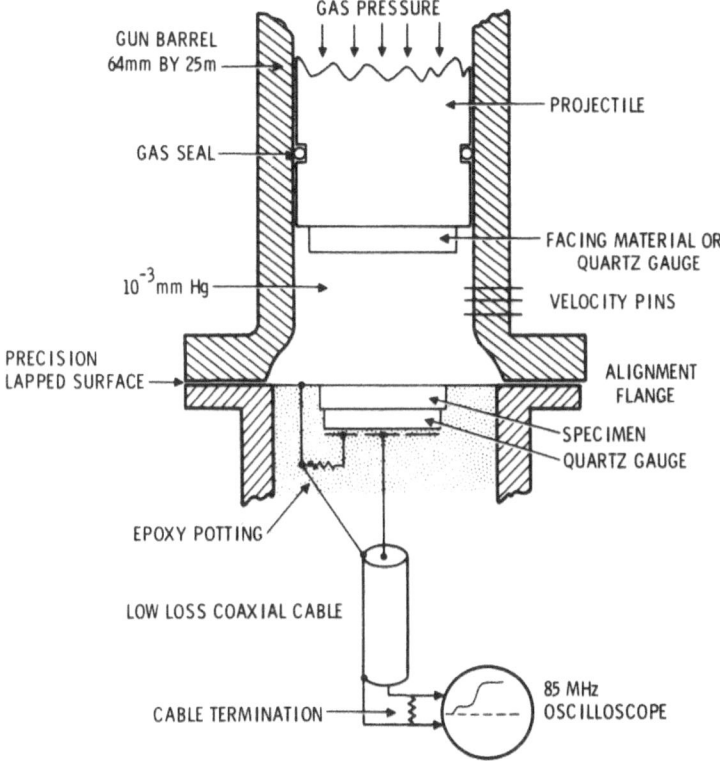

Fig. 5.11. The study of shock compressibility of pressure-sensitive magnetic alloys was carried out with the quartz gauge impact technique. Loading was either with the specimen material or a quartz gauge. Resulting stress pulses were recorded with a quartz gauge (after Graham et al. [67G01]).

used for the study. After machining to the desired dimensions, the samples were heat treated to achieve well defined fcc and bcc phases.

Typical stress-time profiles for the various materials (28.5-at. % Ni, fcc and bcc) and various stress regions are shown in Fig. 5.12. The leading part of the profile results from the transition from elastic to plastic deformation. The extraordinarily sharp rise in stress for the second wave in Fig. 5.12(a) and the faster arrival time compared with that in Fig. 5.12(b) is that expected if the input stress is above the transition, whereas the slower rise in Fig. 5.12(b) is that expected if the stress input to the sample is below the transition. The profile in Fig. 5.12(c) for the bcc alloy was obtained for an input particle velocity approximately equal to that in Fig. 5.12(a) for the fcc alloy. The bcc alloy shows a poorly defined precursor region, but, in any event, much faster arrival times are observed for all stress amplitudes, as is indicative of lower compressibility.

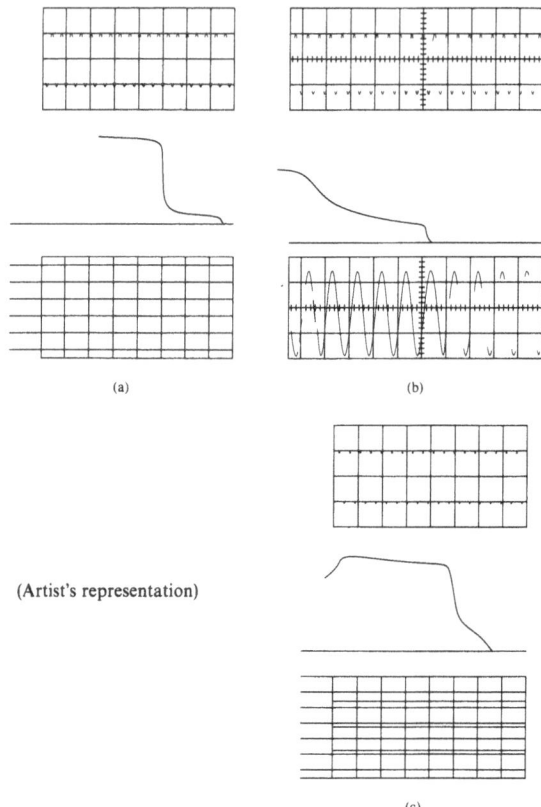

(a)

(b)

(Artist's representation)

(c)

Fig. 5.12. Stress versus time responses (measured currents) for (a) fcc 28.5-at.% Ni above 2.5 GPa (25 kbar), (b) below 2.5 GPa, and (c) for bcc 28.5-at.% Ni, reveal substantial differences in mechanical response. Such records at different input stresses are used to determine the various stress-volume relationships (after Graham et al. [67G01]). Time increases from right to left. Timing waves (upper sinusoidal signal) are 10 MHz.

The excellent time resolution of the quartz gauge has revealed that few materials give idealized step function stress-time profiles. Hence, to reduce the stress-time profiles to stress-volume points the profile observed must be approximated by a series of small incremental steps constructed around the observed profile. This procedure has the further advantage that corrections can be made for the slightly variable response of the gauge. The solution for the incident stress from the measured interface stress was accomplished for each small increment of stress using the linear mismatch relations and the appropriate shock velocity for each step. The resulting final values can be checked against the measured input conditions to the sample as an overall check of experimental errors and errors in data reduction.

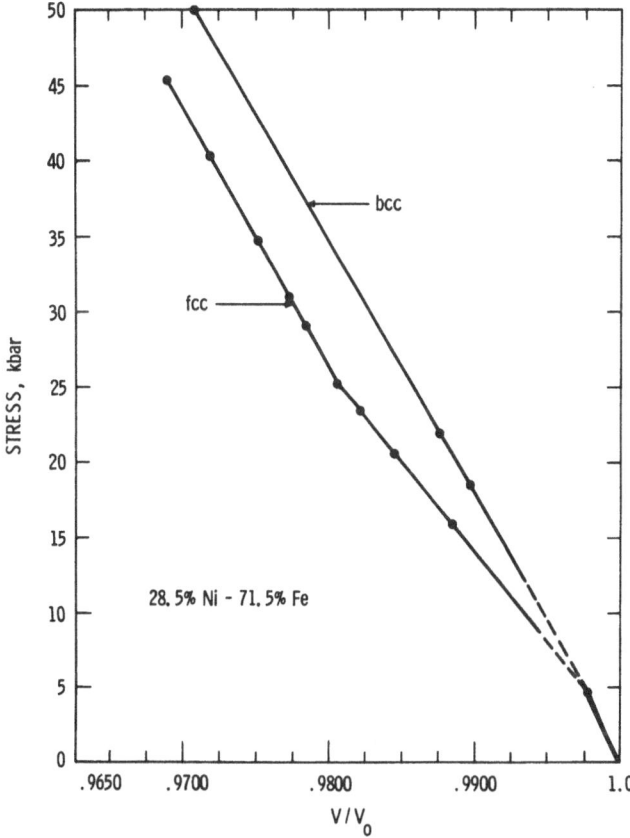

Fig. 5.13. The stress-volume relations of fcc and bcc alloys show the strong compressibility "anomaly" in the fcc phase below 25 kbar (2.5 GPa) associated with the magnetic interactions. Above 25 kbar, the fcc alloy has a "normal" value for compressibility (after Graham et al. [67G01]).

The resulting stress-volume relations for the 28.5-at. % Ni alloys are shown in Figure 5.13. The cusp in the fcc curve at 430 MPa (4.3 kbar) is the mean value observed for the Hugoniot elastic limit, whereas the dashed line shown for the fcc alloy indicates the stress region for which some strain hardening is indicated from the stress profiles. It is readily apparent that below 2.5 GPa (25 kbar) the fcc alloy shows a much larger compressibility than the bcc alloy.

The well defined change in compressibility of the fcc alloy at 2.5 GPa clearly indicates the expected behavior of a second-order phase transition. The anomalously high value of the compressibility for the pressure-sensitive fcc alloy is demonstrated in the comparison of compressibilities of various ferromagnetic iron alloys in Table 5.1. The fcc Ni alloy, as well as the Invar alloy, have compressibilities that are far in excess of the normal values for the

Table 5.1. Compressibilities of nickel-iron alloys (after Graham et al. [67G01]).

Material	Shock compression	Low signal adiabatic
	$(10^{-3}\,\mathrm{GPa^{-1}})$	
Pressure-sensitive alloys		
30 wt% Ni Fe, fcc	8.6 [67G01]	8.97 [67G01]
36 wt% Ni Fe, fcc	8.8 [67G01]	8.9 [67G01]
Pressure-insensitive alloys		
30 wt% Ni Fe, bcc	6.0 [67G01]	6.6 [67G01]
Armco Fe, bcc	6.4	5.93
30 wt% Ni, Fe, fcc above transition stress	5.8 [67G01]	···
36 wt% Ni, Fe, fcc above transition stress	5.0	···
30 wt% Ni, Fe, elevated temp.	6.4 [67G01]	···

ferromagnetic iron alloys whose magnetic properties are not pressure sensitive. Furthermore, for stress in excess of the stress required to induce the transition to the nonmagnetic state, the compressibilities of both 28.5% Ni and Invar have a normal value. Invar does not show a well defined transition, but its roughly analogous behavior to that of the fcc 28.5% Ni alloy shows that magnetic effects are responsible for the pressure-induced change in compressibility.

To further clarify the role of magnetic effects on compressibility, a shock compression experiment was performed on an fcc 28.5-at. % Ni sample whose initial temperature was raised to 130 °C. As is shown in Table 5.1, the compressibility was found to decrease to a value consistent with the nonmagnetic compressibility. Thus, the sharp change in compressibility, the critical values for the transition, and the magnitudes of the compressibility under the various conditions give a clear demonstration that a second-order magnetic transition has been observed, and we will proceed with a quantitative analysis of the transition.

The experiments result in an explicit measure of the change in the shock-wave compressibility which occurs at 2.5 GPa. For the small compressions involved (2% at 2.5 GPa), the shock-wave compression is adiabatic to a very close approximation. Thus, the isothermal compressibility Δk_T can be computed from the thermodynamic relation between adiabatic and isothermal compressibilities. Furthermore, from the pressure and temperature of the transition, the coefficient $d\theta/dP$ can be computed. The evaluation of both Δk_T and $d\theta/dP$ allow the change in thermal expansion and specific heat to be computed from Eq. (5.8) and (5.9), and a complete description of the properties of the transition is then obtained.

The temperature at the observed transition is the initial temperature of the sample added to the shock-compression heating, which is only 3 °C. Uncertainties in the change in Curie temperature are principally due to the measurement of Curie temperature at atmospheric pressure, which was found to be 155 °C. Thus, the Curie temperature is lowered from 155 °C to 25 °C due to the stress of 2.5 GPa.

The small but significant component of shear stress that is associated with the elastic compression results in a stress configuration that is not hydrostatic, and the shock-wave experiment measures a longitudinal component of stress rather than the pressure of the transition. Belov has shown that shear stress does not change the Curie temperature or saturation magnetization of these alloys. His values for $d\theta/dV$ measured in uniaxial stress (which results in a large shear stress) are the same as those obtained hydrostatically. Thus, it is the volume, $0.9807V_0$, that is characteristic of the transition. To compare the present values to the previous hydrostatic pressure values, an equivalent pressure must be computed from our observed value for the volume at the transition. The equivalent pressure is computed from our measured compressibility and the volume change to induce the transition. This yields a value of 2.26 GPa for the equivalent pressure of the transition. From this value of the equivalent pressure and the temperature change induced by this pressure, the value of $d\theta/dP$ is calculated to be -58 °C GPa^{-1}. The present value represents an average over the entire pressure range. As shown in Table 5.2, this value is in good agreement with the results obtained by most of the previous investigators. Therefore, we conclude that $d\theta/dP$ is constant over the entire pressure range encountered.

The thermodynamic description of the transition can now be completed since we now have a measure of both Δk_T and $d\theta/dP$ and can calculate $\Delta\beta$ and ΔC_p from Eqs. (5.8) and (5.9) with results as summarized in Table 5.2. As indicated, the present experiments provide a complete description of the thermodynamic properties of the transition.

The value of the change in thermal expansion coefficient accompanying the transition is considerably larger than that obtained when the transition is thermally induced at atmospheric pressure. The present value for the thermal expansion at 22 °C and atmospheric pressure is 3.0×10^{-5} °C^{-1} and for tem-

Table 5.2. Thermodynamic properties of the shock-compression induced Curie point transition (after Graham et al. [67G01]).

Transition volume	$0.9807\,V_0$
Temperature	26 °C
Equivalent hydrostatic pressure	2.26 GPa
Pressure dependence of Curie temperature	-58 °C GPa^{-1}
Change in compressibility	-2.65×10^{-3} GPa^{-1}
Change in specific heat	-7.2×10^{-3} cal g^{-1} °C^{-1}
Change in thermal expansion	$+4.6 \times 10^{-5}$ °C^{-1}

Compressibility and pressure dependence of Curie temperature are directly measured; changes in specific heat and thermal expansion are calculated from the Ehrenfest relation.

peratures above the transition temperature the value is $5.1 \times 10^{-5}\,°C^{-1}$, which is a normal paramagnetic value for alloys in this composition range. Since the change in thermal expansion shown in Table 5.2 is $+4.7 \times 10^{-5}\,°C^{-1}$, a normal value for the thermal expansion in the high-pressure paramagnetic state implies that in the low-pressure ferromagnetic state the thermal expansion coefficient decreases strongly with pressure to a value close to zero immediately before the transition.

Although there are no direct measurements of the thermal expansion coefficient of this alloy at various pressures, compressibility versus temperature measurements have been made. From these compressibility data and thermodynamic identity $\partial \beta / \partial P = -\partial k / \partial T$ the initial slope of the thermal-expansion-pressure relation can be computed at atmospheric pressure. This initial slope is found to be $+1.7 \times 10^{-5}\,°C^{-1}\,GPa^{-1}$, which is in contradiction to the behavior we have inferred from our high-pressure measurements. However, the extrapolation of initial slopes at atmospheric pressure to high pressures where there are large changes in the magnetic interactions is clearly an uncertain procedure. Thus, the most likely behavior of the thermal expansion coefficient with pressure is an initial small increase followed by a continual decrease in slope until a large negative slope is obtained.

The physical description of strongly pressure dependent magnetic properties is the object of considerable study. Edwards and Bartel [74E01] have performed the more recent physical evaluation of strong pressure and composition dependence of magnetization in their work on cobalt and manganese substituted "invars." Their work contrasts models based on a localized-electron model with a modified Zener model in which both localized- and itinerant-electron effects are incorporated in a unified model. Their work favors the latter model.

The work on iron-nickel alloys has described shock-compression measurements of the compressibility of fcc 28.5-at.% Ni Fe that show a well defined, pressure-induced, second-order ferromagnetic to paramagnetic transition. From these measurements, a complete description is obtained of the thermodynamic variables that change at the transition. The results provide a more complete description of the thermodynamic effects of the change in the magnetic interactions with pressure than has been previously available. The work demonstrates how shock compression can be used as an explicit, quantitative tool for the study of pressure sensitive magnetic interactions.

Direct Magnetization Change Measurement

The magnetization changes accompanying high pressure shock-compression loading can be measured in a relative sense by the use of magnetic "cores" on which excitation and receiving inductive coils have been placed [68G03]. Such an arrangement is shown schematically in Fig. 5.14. The magnetic core is typically composed of a commercial, dense wrapping of 0.15-mm-thick ferromagnetic metal foil, 16 mm wide, wrapped into a rectangular cross sec-

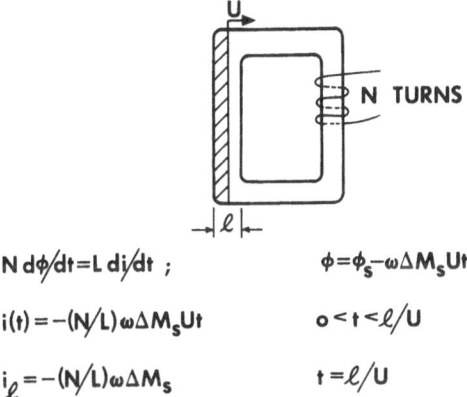

$$N \, d\phi/dt = L \, di/dt \; ; \qquad\qquad \phi = \phi_s - \omega \Delta M_s U t$$

$$i(t) = -(N/L) \, \omega \Delta M_s U t \qquad\qquad o < t < \ell/U$$

$$i_\ell = -(N/L) \omega \Delta M_s \qquad\qquad t = \ell/U$$

Fig. 5.14. A shock wave is depicted propagating at wavespeed U through one leg of a magnetic core with thickness l. The inductance is L, the magnetic flux is Φ, w is the thickness of the core, and the change in magnetization is ΔM. The predicted current-time pulse $i(t)$ is linearly increasing.

tion 5 mm in thickness with a flat surface that can be exposed to direct impact. The thin foils are necessary to minimize eddy currents that could become prominent with rapid changes in magnetization.

On the leg of the core remote from the impact surface, an excitation coil is placed to excite the ferromagnetic sample to a high value of magnetization. In typical situations this magnetization is not at saturation levels. On this same leg, a few-turn copper wire coil is placed to detect the change in magnetic flux accompanying the shock pulse. Such an experiment does not provide a precise value for the absolute magnetization change of the event due to geometric effects on both magnetization and shock conditions, but relative data on maximum measured signal yield coefficients of pressure dependency.

When a toroidal ferromagnetic sample is subjected to shock loading, a pressure wave of pressure P moves through the sample with a velocity U and produces a change in magnetization ΔM. An N-turn detection coil with inductance L is wound around the sample and connected to a resistive circuit in which the L/R time constant is longer than the time required for the shock wave to traverse the sample thickness. The current i in the coil is then

$$i(t) = w(N/L)(\Delta M \, U)t, \quad 0 < t < l/U, \tag{5.10}$$

where t is time, ΔM is the change in magnetization, and w is the sample width. In this situation, the current is proportional to the wavespeed U and change in magnetization ΔM and, when $t = l/U$, the current is a direct measure of ΔM. A typical measured current-time waveform on the commercial Invar alloy is shown in Fig. 5.15.

A plot of relative peak current versus peak shock pressure is shown in

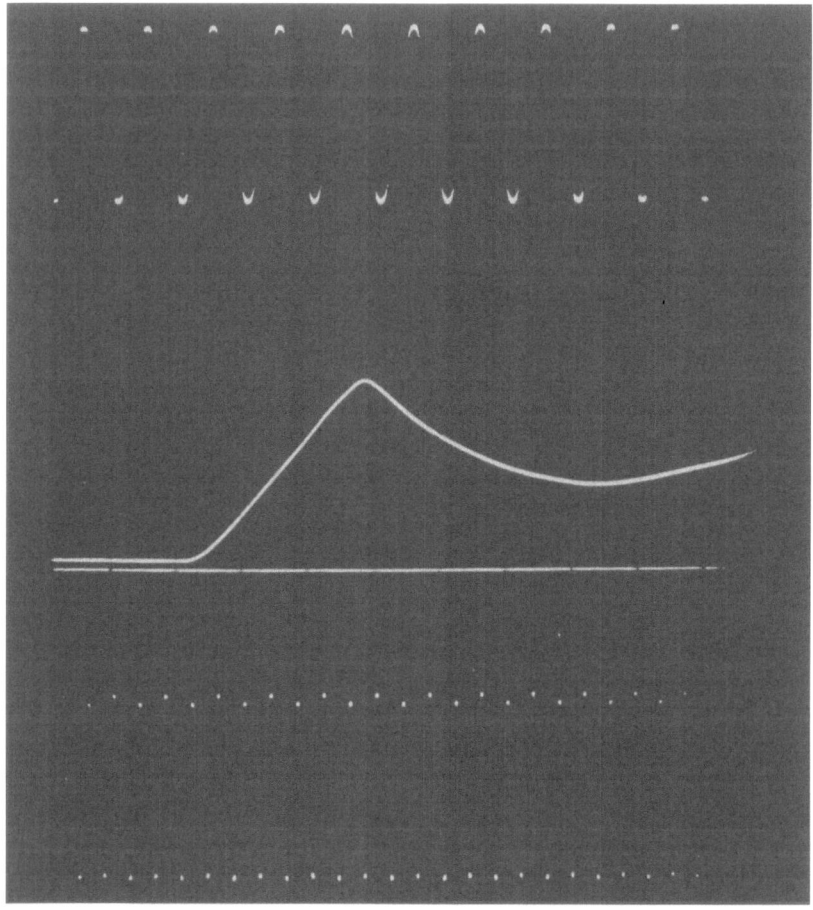

Fig. 5.15. A measured current-time pulse for shock-loaded Invar is shown. Time increases from left to right. The wave shape is closely predicted by the simple theory. Time from impact to peak current is about 1 μs.

Fig. 5.16 for the Invar alloy and other alloys. The behavior is that expected from a Curie point transition with the data indicating a linear magnetization change with pressure. Distinct lower pressure offsets from zero shown in the data may be the result of either or both effects from a nonideal geometric arrangement, or strength or geometric effects in the core. In any event the data provide values for pressure dependence of magnetization that are in good agreement with prior work under static pressure. Wayne has shown that the arrangement of Fig. 5.14 is subject to magnetic form factor effects that affect the absolute magnitude of the signals but are not expected to affect relative measurements at various pressures [69W01].

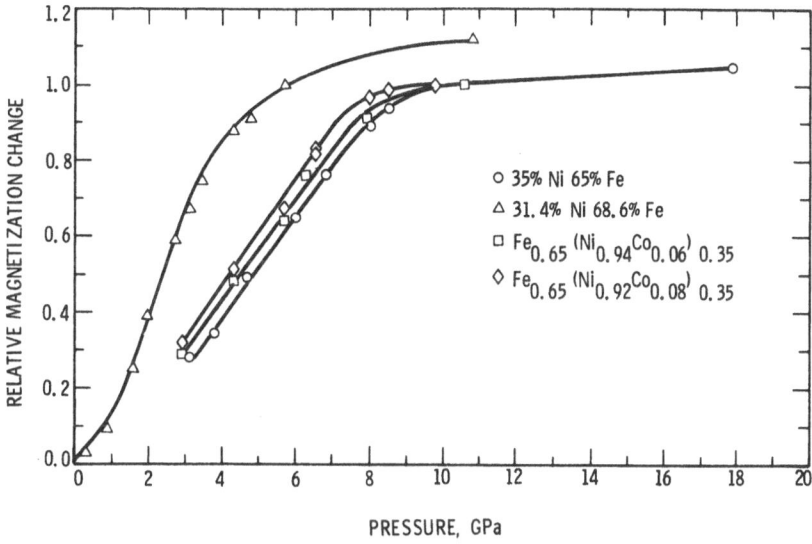

Fig. 5.16. The relative, shock-induced magnetization change is determined at a given pressure by the ratio of peak current to that at full magnetization change. Various sensitivities with pressure are indicated in agreement with static high pressure data. "Offsets" at zero magnetization change are typical and may be due to magnetic or mechanical effects (after Edwards [90E01]).

Shock-Compression Induced Structural Transformations

The shock-compression induced structural phase transformation in iron from the low pressure bcc phase to the high pressure hcp phase is one of the most visible problems studied in shock-compression science, and its discovery was responsible for widespread recognition of the capabilities of the high pressure shock-compression experiment. The properties of many shock-induced phase transitions are summarized in Duvall and Graham [77D01].

When solids experience shock-induced first-order phase transformations, multiple wave fronts propagate through the samples. The leading elastic wave is typically a few GPa in amplitude. The transition wave propagates with the transition pressure, and the trailing wave at the loading pressure. Consideration of the various wave interactions within samples is critical to the interpretation of magnetization change measurements.

The experimental arrangement for direct measure of shock-induced magnetization changes shown in the previous section has been used to study a commercially available iron-silicon alloy, Silectron. The data on the relative amplitude of the peak current versus shock pressure is shown in Fig. 5.17. The features shown are a low pressure region in which little change is observed. At a pressure of about 15 GPa (150 kbar) the signal increases to significantly higher values, finally saturating at a peak pressure of 37.5 GPa

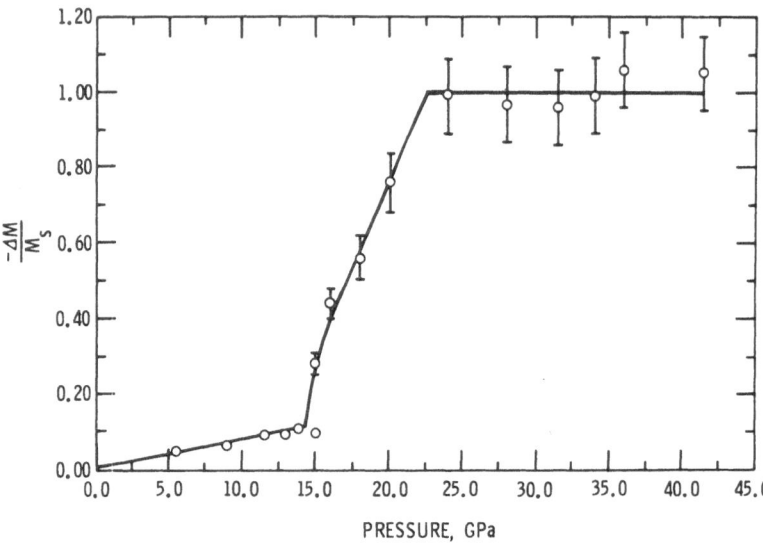

Fig. 5.17. The relative change in magnetization for a 3% silicon-iron alloy shows clear indications for a transition at 14 GPa, the end of the mixed phase region, 22.5 GPa, and the "overdrive" pressure at the Rayleigh line at 37.5 GPa (after Duvall and Graham [77D01]).

(375 kbar). There is a significant change-in slope of the signal at a pressure of about 22.5 GPa, which is interpreted as the pressure to achieve complete magnetization change to a higher pressure state. Between 22.5 and 37.5 GPa there are small increases in signal due to the multiple wave structure within the sample. In this same pressure range there are significant changes in the measured rate of change of current, providing a more definitive measure of wave propagation characteristics within the samples. The observed behaviors indicate that the peak value of current at 37.5 GPa is associated with achieving a pressure sufficiently high to enter the strong shock region.

The indicated transition pressure of 15 GPa is in agreement with the published data with shock-wave structure measurements on a 3% silicon-iron alloy, the nominal composition of Silectron. A mixed phase region from 15 to 22.5 GPa appears quite reasonable based on shock pressure-volume data. Thus, the direct measure of magnetization appears to offer a sensitive measure of characteristics of shock-induced, first-order phase transitions involving a change in magnetization.

5.5 Resistance of Metals

Resistance measurements on substances under high pressure have long been a primary tool to investigate phase transformations and to study piezoresistive effects. Indeed, resistance measurements were the primary diagnostic tool used by Bridgman in his pioneering static high-pressure studies. It is widely recognized that plastic deformation and the defects associated with the deformation have a major effect on observed resistance changes [73L02]. Even though the extensive plastic deformation experienced by metals under shock compression is a serious complication, such measurements are certainly a significant probe of shock-compressed matter.

Numerous resistance measurements have been carried out under high-pressure shock compression [79D01]. Most of the work has been motivated by the desire to develop stress gauges to measure pressures in shock-compressed materials. Other measurements were undertaken to determine critical pressures to induce phase transformations. Although most of the work is not carried out in sufficient detail to relate resistance observations to defect characterizations, excess resistance at given shock pressures is observed in every case compared to comparably loaded static pressure observations. The presence of "residual" resistance for times after the loading is removed provides explicit evidence for irreversible changes in resistance due to defects.

The work of Dick and Styris [75D01] is sufficiently detailed to establish values for vacancy concentrations in silver samples in which resistance measurements were taken. In this work, 15 to 25 μm-thick silver foils were

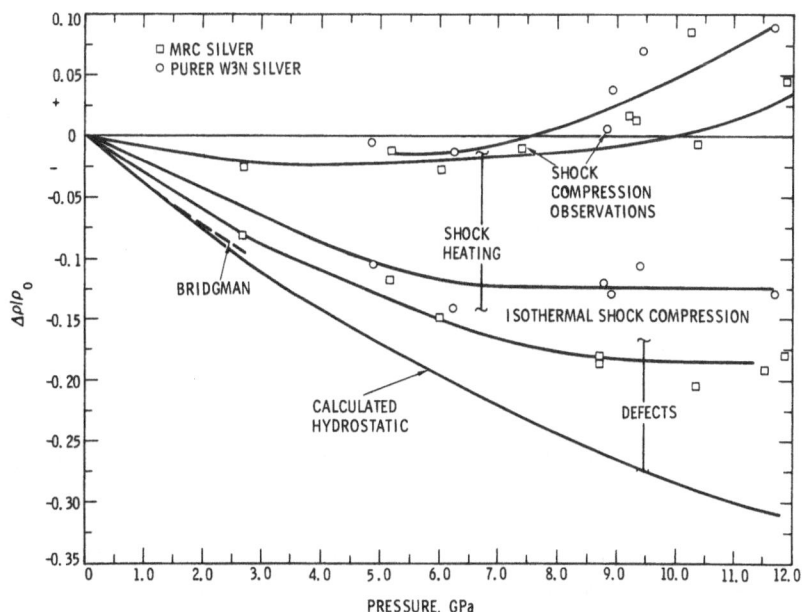

Fig. 5.18. The resistivity of shock-compressed silver foils in excess of that caused by pressure is shown as a function of shock stress. This excess is thought to be due to shock-induced concentrations of point defects ([75D01], after Graham [81G02]).

subjected to controlled shock compression at stresses from 2.5 to 12 GPa. Materials of two characteristically different purities, as indicated by both resistivity measurements at liquid helium temperature and spectroscopic analysis, were investigated. The foil samples were carefully implanted in sapphire disks for careful mechanical impedance matching, and shock loading was carried out with a compressed-gas gun. After accounting for the resistivity change due to the shock-induced rise in temperature and that due to isothermal, hydrostatic compression, an excess resistivity attributed principally to point defects was obtained. The results, shown in Fig. 5.18 as a plot of excess resistivity versus shock pressure for silver in three different starting defect states, show the shock-induced changes to be a strong function of shock pressure. The pure silver and MRC silver were found to exhibit somewhat different behaviors, but all results could be described by a defect resistivity that varied at the $\frac{3}{2}$ power of the compression. The data are well described by peak vacancy concentrations of about 2×10^{-3} observed at 12 GPa.

Gupta and his students have developed procedures for determining the elastic and plastic contributions to shock-deformed metals. The work explicitly recognizes that the metal sample is an inclusion in a host material which may act to cause local deformation unique to the particular host [83G01, 87G01].

5.6 Shock-Induced Electrical Polarization

Shock-induced electrical polarizations can be caused by a number of different physical phenomena. Perhaps one of the most interesting physical observations on shock-compressed matter is the rather ubiquitous self-generated electrical polarization signals from dielectric materials caused by essentially unknown physical processes. These signals have characteristics representative of volumetric electrical polarizations. In most cases, they are thought to be unique to the shock-compression process. Their existence provides overt evidence for nonequilibrium processes under shock compression. Although there are alternate explanations for the underlying shock processes, the influence of shock-induced defects is certainly a major consideration. Generally, two characteristic classes of materials are seen to exhibit the polarizations: ionic crystals and polymers.

It is of interest to compare the observations with different physical mechanisms as shown in Fig. 5.19. Typically, the polarization values for polymers are weak and do not overlap those of piezoelectrics. What is observed is that there is over a 6 order-of-magnitude range in polarizations from the weakest signals (Teflon) to the strongest (PZT 95–5). The polarization signals from ionic crystals are stronger than those in polymers and overlap those of piezoelectrics, albeit at larger strains.

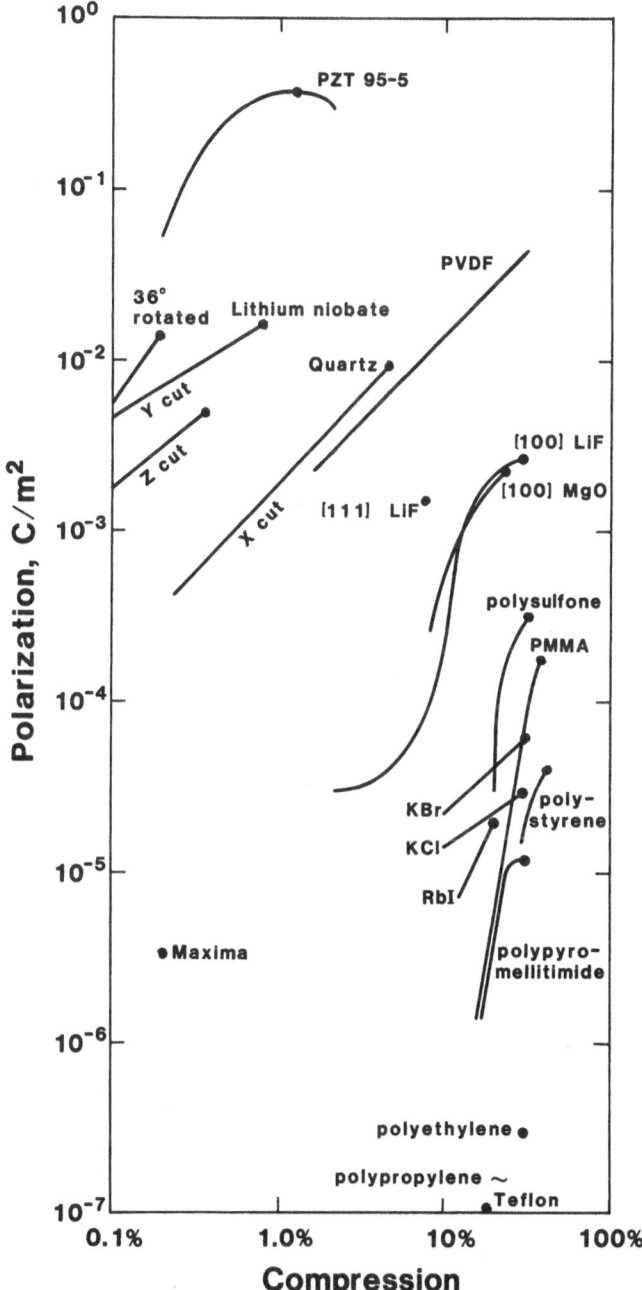

Fig. 5.19. Shock-induced volume polarizations have been observed in a wide range of solids due to a number of different physical phenomena, including piezoelectricity and ferroelectricity. The signals observed from polymers and ionic crystals are not due to established phenomena, and are described as due to shock-induced polarization effects.

Shock-Induced Polarization in Ionic Crystals

Numerous observations of the effect in ionic crystals were carried out by Mineev and Ivanov in the Soviet Union [76M01]. This is a class of crystals in which a number of materials factors can be confidently varied. By choice of crystallographic orientation, various slip directions can be invoked. By choice of various crystals other physical factors such as dielectric constant, ionic radius, and an electronic factor thought to be representative of dielec-

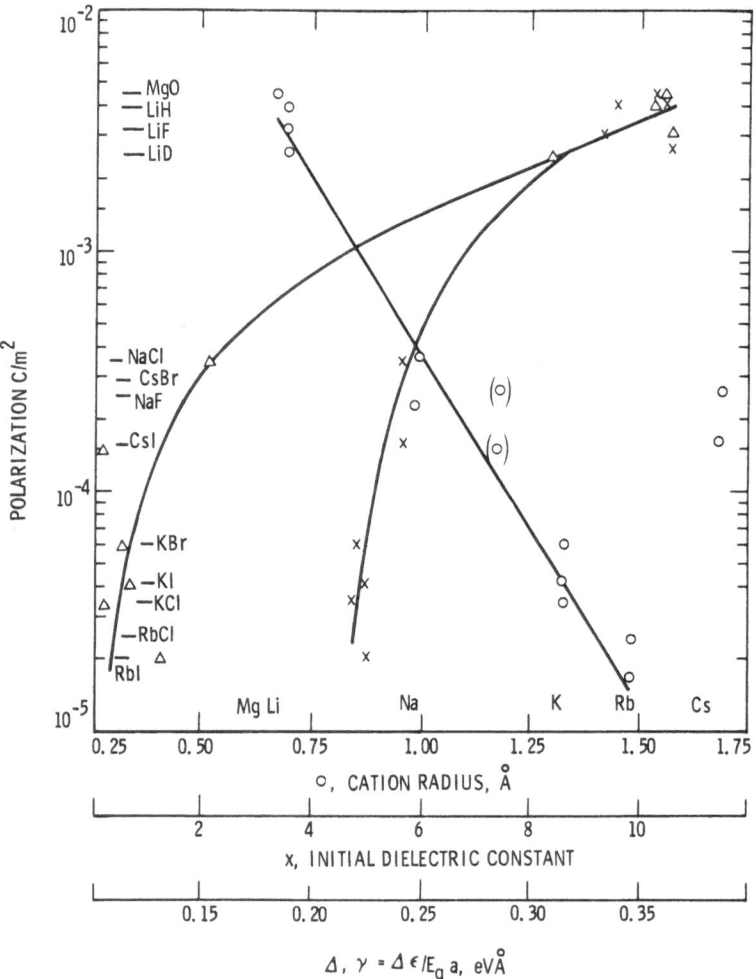

Fig. 5.20. The shock-induced polarization of a range of ionic crystals is shown at a compression of about 30%. This maximum value is well correlated with cation radius, dielectric constant, and a factor thought to represent dielectric strength. A mechanically induced point defect generation and migration model is preferred for the effect (after Davison and Graham [79D01]).

tric breakdown can be varied. The summary of these investigations is shown in Fig. 5.20.

Based on their numerous investigations, Mineev and Ivanov concluded that, at compressions less than about 30%, the overall features of shock-induced polarization in ionic crystals can be semiquantitatively described by the existence of cation-vacancy dipoles. These dipoles are thought to result from shock-induced generation of large numbers of point defects and subsequent displacement of the cation over a distance of 1–10 lattice parameters in times of about 50 ns. during the "rise time" of the loading pulse. Defect concentrations of 10^{23} m^{-3} at compressions of 10%, increasing by an order of magnitude for each subsequent 10% compression, are found to be compatible with the observations. Relaxation of the polarizations is thought to follow thermal equilibration of the shock-induced defect structure.

Shock-Induced Polarization of Polymers

Although signals are typically more feeble than for ionic crystals, shock-induced polarizations are widely observed in polymeric materials (See Graham [79G01, 82G02]). The experimental arrangement of the studies, and a typical record, is shown in Fig. 5.21. The various observations have been collected and extended to a number of interesting polymers with polarization values as summarized in Fig. 5.22. In the figure, the "effective polarization" (a represen-

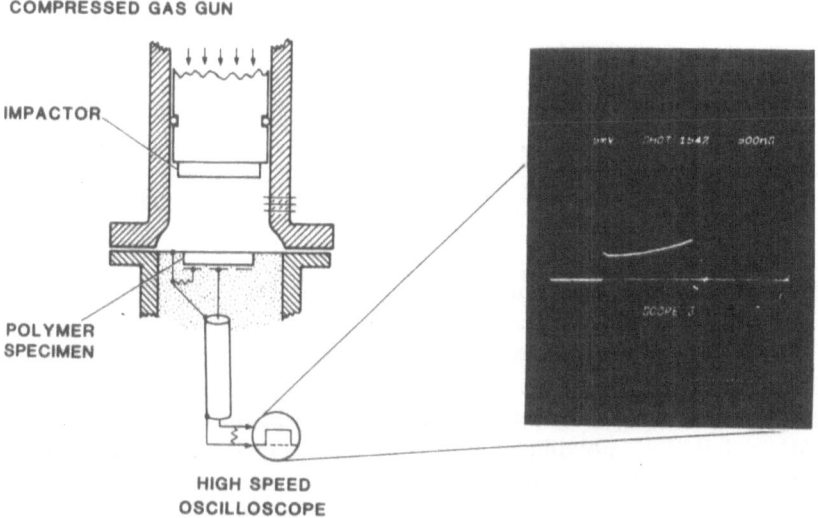

Fig. 5.21. The shock-induced polarization of polymers as studied under impact loading is shown. For the current pulse shown, time increases from left to right. The increase of current in time is due to finite strain and dielectric constant change. (See Graham [79G01]).

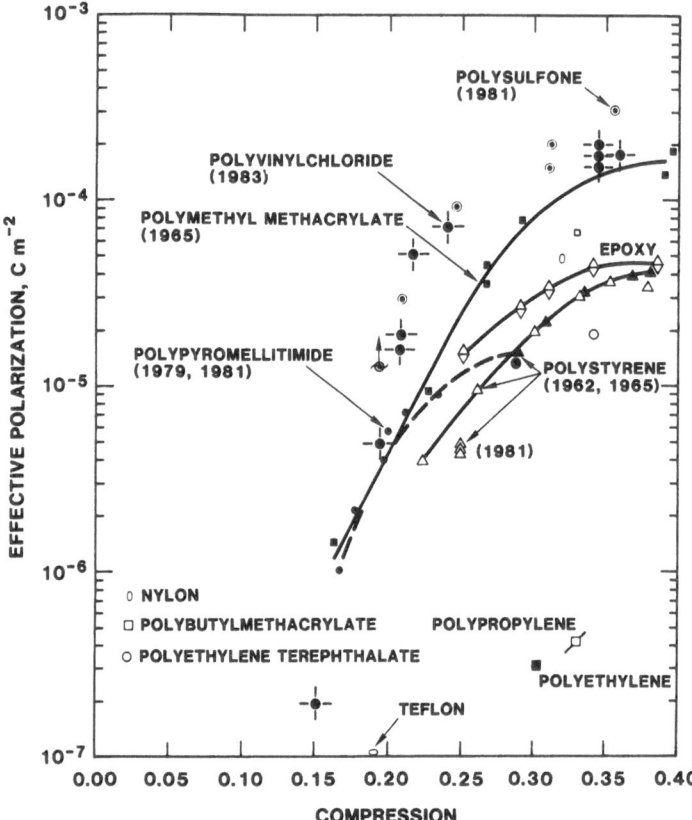

Fig. 5.22. Shock-induced polarization of polymers has been studied by many investigators, with data as summarized. The typical behavior indicates a threshold compression of about 10%–15% followed by a rapid increase in value. The polarizations shown vary over three orders of magnitude. The author has proposed a mechanically induced bond-scission model to describe the effects. (See Graham [79G01].)

tation that is independent of assumptions of dielectric constant) is shown as a function of the compression.

Several overall features of the shock-induced polarization are apparent. First, there appears to be a threshold compression below which the signals are not observed. The compression for this threshold is considerable, about 15%, such that it is not difficult to believe that the material must be considerably altered in structure before polarizations appear (shown in Fig. 5.22). Following the threshold compression, the polarizations increase extraordinarily rapidly with increasing compression, finally reaching a saturation value at compressions of perhaps 30%.

When the magnitudes of the polarizations are compared to the structure of the monomer, a striking correlation exists between the structure and the

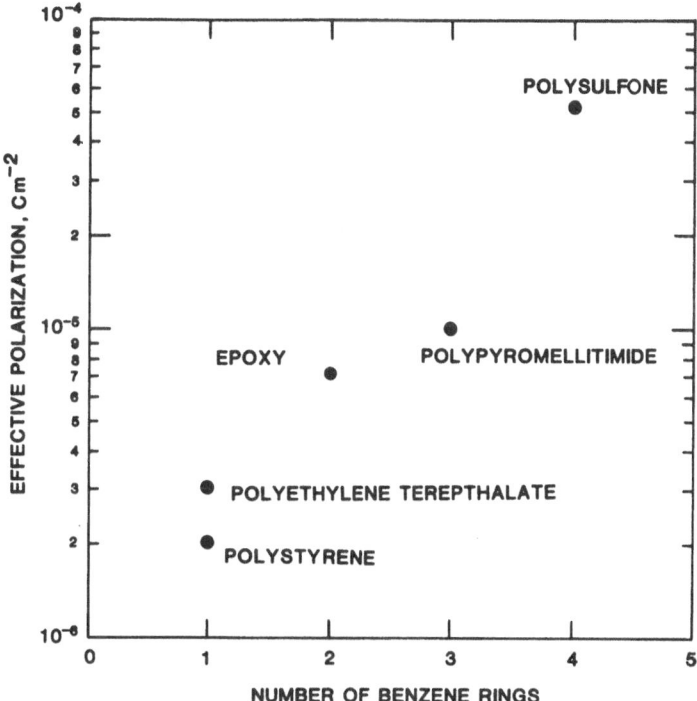

Fig. 5.23. The shock-induced polarization of the indicated polymers containing benzene rings in their structure is shown to be strongly dependent on the number of rings (after Graham [82G02]).

magnitude of the polarization; the more complex structures show the largest polarizations. Figure 5.23 shows the relation between the number of benzene rings in the structure and the observed maximum electrical polarization. The simple structures of polyethylene and Teflon show only the faintest evidence for a polarization. It is also observed that the same polymers that exhibit strong polarization are those that become the most electrically conductive under shock compression [79G01].

These observations were the basis for the proposal that polymers, like ionic crystals, exhibit shock-induced polarization due to mechanically induced defects which are forced into polar configurations with the large acceleration forces within the loading portion of the shock pulse. Such a process was termed a "mechanically induced, bond-scission" model [79G01] and is somewhat supported by independent observations of the propensity of polymers to be damaged by more conventional mechanical deformation processes. As in the ionic crystals, the mechanically induced, bond-scission model is an example of a catastrophic shock compression model.

Alternately, a benign shock compression model has been invoked to explain the polarizations. An elastic dipole-rotation model was first proposed by Hauver to describe the effect. In this case, polar electrical structures in the undamaged structure are rotated to new positions in the acceleration field of the loading pulse. Although such a model appeals to established physical processes, it fails to describe the observations. Such a model has no threshold, and relaxation times of the dipole-rotation model will be subnanosecond, in disagreement with the observations. The contrast between the elastic-dipole-rotation model and the mechanically induced bond-scission model bring catastrophic- and benign-shock paradigms into sharp contrast. Although it is difficult to identify observations at the continuum level with molecular level processes, clearly the more realistic conditions of catastrophic shock compression must be given serious consideration.

5.7 Electrochemistry

In Part IV of this book, studies of physical and chemical changes in highly porous powder compacts of inorganic solids will be described. The technology developed for the porous solid studies has been applied to the study of electrochemistry under high pressure shock compression [88G02]. In the electrochemistry work, galvanic cells with porous electrolytes of various porosities are subjected to controlled shock loading and their electrical responses are measured while under shock loading and after the pressure is released to atmospheric values. The work also includes the influence of external electrical loads on the galvanic cell response. In the electrochemical cell problem, the effect is controlled to first order by melting of the solid electrolyte. In principle, the measured open circuit voltages provide a measure of electrochemical potentials at high pressure.

The galvanic cell studied (shown in Fig. 5.24) utilizes a highly porous solid electrolyte that is a eutectic composition of LiCl and KCl. This eutectic has a melt temperature of 352 °C and has been carefully studied in prior electrochemical studies. Such solid electrolytes are typical of thermal battery technology in which galvanic cells are inert until the electrolyte is melted. In the present case, shock compression activates the electrolyte by enhanced solid state reactivity and melting. The temperature resulting from the shock compression is controlled by experiments at various electrolyte densities, which were varied from 65% to 12.5% of solid density. The lower densities were achieved by use of microballoons which add little mass to the system but greatly decrease the density.

The cathode is FeS_2 and the anode is a Li(Si) alloy. These materials were also selected as materials that have been carefully studied in thermal battery technology. All materials in the cell are moisture sensitive and were handled under "dry room" conditions available at the author's laboratory.

When lithium ions become sufficiently mobile due to heating, they migrate from the anode to the cathode with the reactions shown in Fig. 5.24 and produce open circuit voltages of about 2.5 V under ideal conditions. In

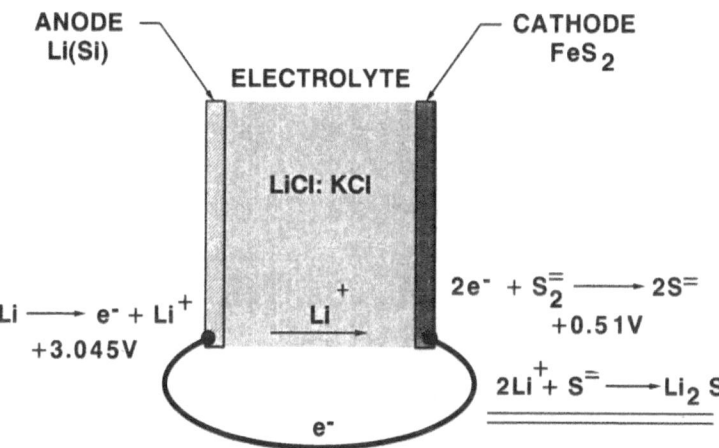

Fig. 5.24. The electrochemical properties of the galvanic cell shown have been studied under high pressure shock compression. The cell is composed of anode, electrolyte, and cathode materials studied in independent applications of thermal batteries.

the present case, there is concern for introduction of physical and chemical changes in the porous anode, electrolyte, and cathode.

For fixed peak shock pressure the measured currents typically showed a large signal that reached a peak at the time of maximum pressure. The magnitude of this current was a strong function of the starting density of the electrolyte. Upon release of pressure, the currents reached a higher value of current. Both the influence of electrolyte density and the higher signals upon release of pressure are consistent with the known melting behavior of the electrolyte. Figure 5.25 shows pressure-temperature paths for the various loadings and an estimate of the melt curve of the electrolyte based on measurements on LiCl under static pressure. In the figure, the shock temperatures shown are controlled by the porosity of the electrolyte.

Experiments on the influence of electrical resistance on voltages produced by the cell show that at resistances greater than 1.5 V the signals approach the predicted galvanic potential of 2.5 V. At resistances of 0.1 Ω, the signals under pressure and upon release of pressure are close to the same magnitude. The indicated internal resistance of the cell gives resistivity values significantly greater than those of conventionally melted LiCl:KCl. Apparently, the shock-melted electrolyte is incompletely melted, possibly due to nonequilibrium thermodynamic, heterogeneous conditions.

The study clearly shows that the observed electrical signals are electrochemical in origin, and the first-order description of the process is consistent with that expected from atmospheric pressure behaviors. Nevertheless, the complications introduced by the shock compression do not permit definitive conclusions on values of electrochemical potentials without considerable additional work.

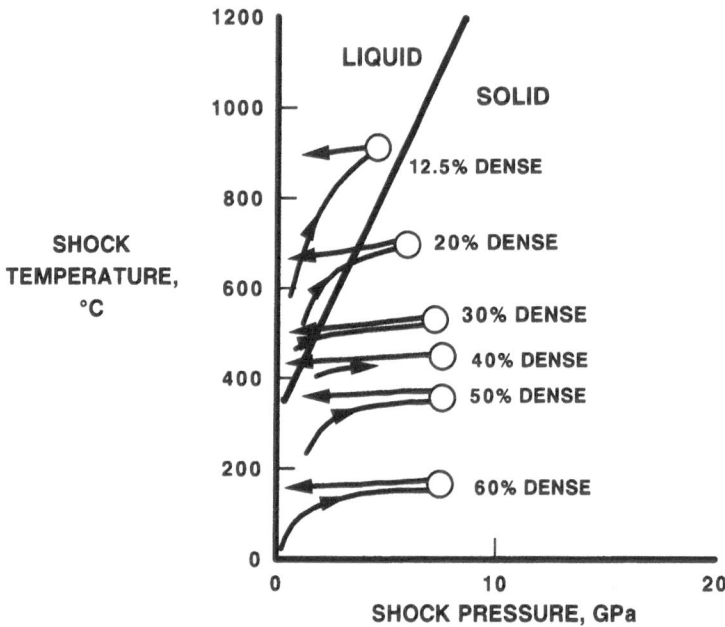

Fig. 5.25. The shock temperature in LiCl:KCl electrolytes is controlled with the use of electrolytes with initial densities as shown. The circle represents the shock conditions. Upon release of pressure the final temperature is expected to cross the melt curve for certain initial conditions.

5.8 Elastic-Plastic Physical Properties

In this chapter studies of physical effects within the elastic deformation range were extended into stress regions where there are substantial contributions to physical processes from both elastic and inelastic deformation. Those studies include the piezoelectric responses of the piezoelectric crystals, quartz and lithium niobate, similar work on the piezoelectric polymer PVDF, ferroelectric solids, and ferromagnetic alloys which exhibit second- and first-order phase transformations. The resistance of metals has been investigated along with the distinctive shock phenomenon, shock-induced polarization.

The wide range of these physical effects provides an unusual insight into shock-compressed matter not possible from conventional measurements of wave profiles or conventional "Hugoniot" measurements. Indeed, the picture that results is considerably more complex than revealed in the conventional mechanical measurements and gives little confidence in the benign shock-compression assumptions, which neglect the role of defects. In most cases the physical picture that emerges from the various studies is the major role that shock-induced defects play in controlling physical properties. Questions concerning the quantitative aspects of defect generation and motion are not directly treated in mechanical measurements. Only in very specialized cases do measurements of physical properties reveal quantitative aspects of defects. Such detailed information has typically come from samples that have been subjected to controlled shock compression and preserved for post-shock analysis.

With the extensive background established for piezoelectric responses of shock-compressed crystals in the elastic range, higher pressure studies can confidently be interpreted. It has been found that the three-zone model of Neilson and Benedick has appropriate first-order features to interpret many of the observed elastic-inelastic responses. The electrical studies were used to determine mechanical states and provide a completely independent determination of the reduction of shear strength observations from wave profile studies. Indeed, it should be recognized that the first proposal for such a materials behavior was from the piezoelectric studies of Neilson and Benedick. The electrical measurements provide special insight into the reduction in strength with evidence for finite delays in the process as well as evidence for a small, but finite, residual strength. Given the complexity of the process, the electrical observations are of particular value. Particularly striking is the observation of shock-induced conduction in crystals such as quartz and lithium niobate. These processes are apparently controlled by shock-induced defects. Although the three-zone model of shock-compressed crystals was found to describe the behavior of x-cut quartz, it failed to describe observations on lithium niobate except in special circumstances.

The work on piezoelectic polymers is especially significant technologically. The original work on the quartz gauge led the way for a new generation of time-resolved stress gauges, but application was limited to such low stresses that its influence was limited. The higher operating stress ranges of the piezoelectric polymers and unobtrusive nature of the thin film PVDF make possible a range of measurements not previously possible. The ability to provide a direct measure of *stress rate* opens a new set of phenomena for detailed study.

The piezoelectric polymer investigations give new physical insight into the nature of the physical process in this class of ferroelectric polymers. The strong nonlinearities in polarization with stress are apparently more a representation of nonlinear compressibility than nonlinear electrical effects. Piezoelectric polarization appears to be linear with stress to volume compressions of tens of percent. The combination of past work on PVDF and future work on copolymers, that have quite different physical features promises to provide an unusually detailed study of such polymers under very large compression.

Studies of the electrical and mechanical responses of ferroelectric solids under shock compression show this technical problem to be the most complex of any investigated. The combination of rate-dependent mechanical and electrical processes along with strong electromechanical coupling has clouded physical interpretation of the numerous investigations.

The work on ferromagnetic alloys principally demonstrates that shock compression can be used to study magnetic behavior at pressures higher than that of conventional static high pressure experiments. These magnetic properties are not sensitive to defect configurations or concentrations, and there is good agreement between the shock data and that from independent investigations. The second-order, Curie point transition study is particularly distinctive and is an area that could be pursued in some detail in those alloys in which the Curie temperature is sensitive to pressure.

Studies of the resistance of metals provides a stark example of a conceptually simple measurement that is very difficult to execute and correspondingly complex in interpretation. In this phenomenon, the role of shock-induced defects is dominant.

Finally, the phenomenon of shock-induced polarization represents perhaps the most distinctive phenomenon exhibited by shock-compressed matter. The phenomenon has no counterpart under other environments. The delineation of the details of the phenomenon provides an unusual insight into shock-deformation processes in shock-loading fronts. Description of the phenomenon appears to require overt attention to a catastrophic description of shock-compressed matter. In the author's opinion, a study of shock-induced polarization represents perhaps the most intriguing phenomenon observed in the field. In polymers, the author has characterized the effect as an "electrical-to-chemical" investigation [82G02].

Among the newer probes now being developed, spectroscopic observations of crystals in the elastic-plastic regime hold promise for limited development of atomic level physical descriptions of local defects [91S02]. It is yet to be determined how generally this probe can be applied to solids. The electro-chemical probe appears to have considerable potential to describe shock-compressed matter from a radically different perspective.

PART IV

Chemical Processes in Shock-Compressed Solids

CHAPTER 6

Shock-Compression Processes in Solid State Chemistry

In this chapter: The background of shock-induced solid-state chemistry; conceptual models and mathematical models; chemical reactions in shock-compressed porous powders; sample preservation.

6.1 Background

From humble, controversial beginnings, shock-induced solid state chemistry has grown into perhaps the most forward-looking area in shock-compression science. For over thirty years scientific studies of shock-compressed solids were considered the province of physics (at least outside the Soviet Union). Slowly, over the past ten years it is becoming apparent that observations of solid state chemistry pose questions that challenge our descriptions of shock-compressed solids based on observations of solid state mechanical, physical characteristics. The scientific community is presently only beginning to consider certain of the chemical issues, but vigorous activities are beginning to develop a foundation for an improved framework delineating the problem.

It is indeed a distressing prospect to contemplate the complications introduced by chemical changes into an otherwise orderly physical description. The chemical complications are intimately intertwined with the mechanical and physical effects, which are already understood to be more complex than present theory indicates. As the questions addressed in solid state chemistry are quite different from those addressed in prior work, new approaches are required to develop a scientific understanding of the field.

Based largely on arguments involving the benign shock-compression paradigm, chemical effects were not considered to be significant for many years. Given diffusion rates in typical defect-free solids, chemical changes cannot occur over significant volumes in the microsecond duration of the shock-compression event. But, with our present understanding of shock-compressed matter within the catastrophic shock paradigm, it is clear that we should expect shock-compression events to cause significant chemical changes.

Solid state chemistry is controlled by defects, and the highly defective, shock-compressed solid provides the optimal condition for chemical change.

Furthermore, the high kinetic energy associated with shock compression can lead to relative mass motion over distances of many microns in microsecond time durations. Any chemical change further intensifies heterogeneities resulting from plastic deformation. Thermochemical effects can lead to significant localization of thermal energy. All localization phenomena occur within the large kinetic energy flux of the loading. Given these conditions, the question is not whether chemical changes can occur, but the degree of chemical change expected in any situation. These effects are likely limited to areas of local chemical heterogeneity or to regions of macroscropic deformation heterogeneities in fully dense solids. In low density powder compacts, the unoccupied space available for mixing of reactants results in a scale for chemical change that can readily encompass areas millimeters or larger in size in times of 1 μs.

At present, the principal thrust of shock-induced solid state chemistry rests upon preservation of samples that have been subjected to controlled, quantitative, reproducible shock loading. Such an endeavor must, of necessity, involve close, cooperative efforts between materials communities, including solid state physics, materials science, and shock-compression science.

Numerous texts describing conventional solid state chemistry provide necessary background within which the shock processes must be developed. From a historical viewpoint, the monograph of Hedvall [66H01] describes the early origins of solid state chemistry from the time in the early 20th century in which the dominant scientific view was that solids did not react; chemistry must take place in the liquid state. Echoes of this view reverberate in contemporary shock-chemistry communities. A basic text on solid state reactivity is that of Schmalzreid [81S01], who develops a basic picture of the underlying principles. It is significant that the early portions of this text are concerned with descriptions of defects and their influences on solid state chemistry. The text of West [84W01] is a thorough, wide ranging account with many processes and analyses considered.

It is to be recognized that the field of solid state chemistry is qualitatively distinct from liquid or gas-phase chemistry. This distinction results from the limited atomic mobility in solids. Without an ability for ionic species to diffuse over distances large compared to atomic spacing, no significant chemical change can occur. For this reason, solid state chemistry is controlled by defects; lattice structural features of solids principally affect chemistry through their influence on defect structure. The defects act to greatly enhance atomic mobility. In any particular chemical reaction, the morphology of the reactants plays a leading role. This again is due to the limited atomic mobility. Morphology plays a dominant role in the intimacy of the reactants.

The plan of this chapter is first to briefly recall the history of work in solid state chemistry. Following this, the mechanisms that the author proposes control shock-induced solid state chemistry will be considered in terms of shock-induced changes to potential reactants. Enhancements in solid state chemical reactivity are considered in Chap. 7. There are many groups who have made contributions in solid state chemistry whose early work is summarized in a published bibliography [83G02]. Work on chemical processes in geophysical materials has recently been reviewed by Boslough [91B01]. The present volume emphasizes work of the author and his colleagues in order to provide a degree of consistency in experimentation and analysis.

Early Work

Research on shock-induced solid state chemistry stems from worldwide recognition during the decade between 1956 and 1966 that shock compression could be used to subject materials to unique, unexplored processes of extraordinary intensity. This situation came about after scientific principles underlying shock compression were presented to the world by the publication of the 1958 Los Alamos review [58R01]. With these principles established, it became possible to think of the shock process as a controllable event to be exploited in materials synthesis. The most crucial event was the work of DeCarli and co-workers [61D01], who showed that diamond could be synthesized without a catalyst from graphite in a shock process. Given the technological importance of diamond, this observation stimulated a major activity over the next 10 yr. The commercial production of diamond with shock processing is still perhaps the most outstanding example of commercial success in materials synthesis with shock processing [68C01]. With the recent expiration of the DuPont patents in the area, there is considerable activity worldwide to develop commercial diamond synthesis processes.

From work centered at the Japanese Defense Academy, Kimura [63K03], followed by Horiguchi and Nomura, demonstrated the first chemical synthesis from shock-compressed mixed powder porous compacts, as well as shock activation as displayed in enhanced catalytic activity and in enhanced sinterability of shock-modified materials. Batsanov [65B01] demonstrated chemical activity in shock-modified potassium nitrate in 1965 and his publications show that he recognized the significance of "the new chemical trend" of solid state chemistry. Adadurov and co-workers [65A02] reported the shock-induced polymerization of acrylamide and trioxane in that same year. With the demonstration of enhanced sinterability of a range of ceramics by Bergmann and Barrington [66B01] in 1966, all of the principal areas of shock-induced solid state chemistry had been demonstrated in the early period. A summary of these early solid state chemical synthesis activities is shown in Table 6.1.

Solid state chemistry was vigorously pursued in the Soviet Union from their earliest work, but other shock-compression groups showed little interest in the area. Within a benign shock compression picture, such chemical effects could not occur in the microsecond duration of the shock pulse. Observations of chemical changes must therefore be interpreted to be the result of poor experimental control or processes that occurred long after the shock event.

Table 6.1. Early solid state synthesis studies.

Author	Date	Synthesis	Remarks
Parsons [20P01]	1920	graphite to diamond	unsuccessful
Riabinin [56R01]	1956	sublimation, dissociation	
Grover et al. [58G01]	1958	sublimation, dissociation	abstract
DeCarli and Jamieson [61D01]	1961	graphite to diamond	the key work
Milton and DeCarli [63M01]	1961	maskelynite	mineral synthesis
Kimura [63K03]	1963	TiC	from mixed powders
Adadurov et al. [65A02]	1965	polymerization	acrylamide, trioxane
Batsanov et al. [65B01]	1965	potassium nitrate	
Horiguchi and Nomura [65H02]	1965	catalytic activity	carbon
Horiguchi and Nomura [65H02]	1965	shock activation	WC
Bergmann and Barrington [66B01]	1966	shock activation	ceramics

6.2 Conceptual Models

In the early 20th century [66H01], solid state chemistry began to move from a paradigm that "only fluids react" to today's well developed scientific field in which underlying principles are well established. Today, shock-induced solid state chemistry is beginning to move from a benign shock paradigm under which solids cannot react in the microsecond duration of the shock process to a realistic representation within the catastrophic shock process. Today, we move from an "only fluids react" paradigm for shock-induced chemistry to a recognition that melting and liquid states may, in fact, inhibit reactions.

In solid state chemistry the limited atomic mobility in the solid state controls chemical changes and leads to explicit consideration of the relative location of potential reactants (the configuration) and solid state reactivity as controlled by solid state defects. The same factors dominate shock-induced solid state chemistry.

In the shock process the problem is to identify those features that control location of reactants and to quantify the solid state defect properties. The author has proposed that the shock process should explicitly identify the effects in the transition zone of a shock compression pulse that carry the potentially reacting materials from their initial configuration to the high pressure, compressed, high temperature state. These factors are shown in the cartoon of Fig. 6.1, which depicts an initial configuration (shown as parallel plates) being changed to a new shock-compressed configuration.

Although it is probably not possible at present to develop a completely satisfactory quantitative model, a conceptual model that identifies the critical processes can be developed. First, it is apparent that before significant chemical reaction can occur substantial modification must be induced by the shock-compression process. With present knowledge, the problem is one of mechanical deformation, rather than one of chemistry. The materials studies

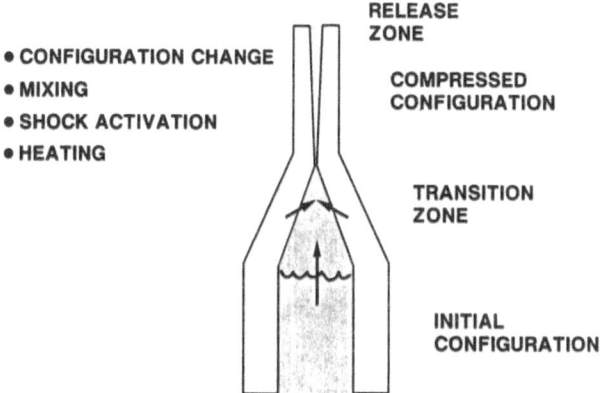

CONFIGURATION CHANGE

MIXING

SHOCK ACTIVATION

HEATING

RELEASE
ZONE

COMPRESSED
CONFIGURATION

TRANSITION
ZONE

INITIAL
CONFIGURATION

Fig. 6.1. The various studies on shock-induced solid state chemistry led to conceptual delineation of the factors most important to the process as shown. In the figure, the configurations are indicated by parallel plates whose spacing is changed by the shock-deformation process. In the transition zone from the initial configuration to the shock-compressed state the indicated factors of configuration change, mechanical mixing, shock activation, and heating operate to produce the conditions controlling the initiation of chemical reaction (after Graham [89G01]).

and knowledge of shock-compression processes can be combined to identify the overall critical features.

The figure characterizes materials in some initial configuration, which is altered in time as a loading pulse sweeps over it. The shock-compression event is characterized by a transition zone in which significant changes are occurring. After the transition, the material is in a substantially different state, and, finally, the pressure is released.

The most critical aspects of the process are those that occur in the transition zone, although it should be recognized that the initial configuration has direct influence on the subsequent processes in the transition zone. In Fig. 6.1, the four critical features are identified as (1) the configuration change; (2) mechanical mixing; (3) shock activation; and (4) heating.

To further characterize the event it is first necessary to identify critical features of the initial configuration that will strongly influence the process. For powder compacts, the most obvious features are the morphological characteristics of the powders, their microstructures, and the porosity of the compact. For solid density samples, the grain structure, grain boundaries, defect level, impurities, and inclusions are critical features.

Due to the shock deformation, the materials configurations are drastically altered. Materials are grossly deformed. Voids are eliminated. The morphologies of all substituents are changed. Any model that attempts to treat the chemical reaction problem must deal with the realistic configuration, not the starting configuration. Determination of the realistic configurations of potential reactants at times of incipient chemical reaction is readily achieved from careful examination of samples that are shocked under controlled loading conditions and preserved for post-shock examination. There are numerous examples of such studies reported in Chap. 7.

The most critical mechanical factor that controls chemical change is the mixing between reactants that are originally in a form in which chemical reaction will not occur. The enormous kinetic energy during the shock process acts to mix the reactants due to localization of the energy. The voids serve principally as a space within which the mixing can occur. From our knowledge of shock conditions, factors influencing the mixing can be identified as the shock impedance of the powders, the difference in impedance of powders, the strength of the powders, and the effective viscosity in viscoplastic high speed flow. Dremin and Breusov [68D01] have emphasized in their "roller model" that mass motion resulting from atomic level defects can greatly speed structural and chemical reactions due to mechanically forced local mass motion.

To the extent that mass motion due to differential material velocity is a significant factor in initiating reaction, it is the volumetric proportions of the reactant mixture that are critical, rather than the molar proportions. Relative motion of the potential reactants required to place them in a more intimate configuration occurs over a limited time, leading to consideration of spatial (volumetric) limitations to initiation of reaction. If reactant densities are significantly different, the volumetric proportions may differ quite significantly from the molar proportions. Experimental evidence shows that volumetric distributions close to one-to-one ratios or 40:60, 60:40 are the most favorable for initiation of reaction.

As shown in Chap. 7, shock compression introduces large numbers of defects which in turn cause substantial increases in solid state reactivity. Such shock activation is obviously critical to the process. One of the most direct effects of the mechanical deformation is the removal of oxides or other surface films from the surfaces of the powders. It is well recognized that such surface films can greatly inhibit chemical reaction. The very large mechanical deformation would be expected to substantially damage, if not completely remove, such films. Other manifestations of shock activation are shown in the next chapter. Effects have been shown that represent many orders of magnitude of change in solid state reactivity.

Finally, there must be heating to achieve the temperature necessary for the reaction to proceed. In most cases in porous solids there are sufficient temperatures produced to provide the proper thermal environment. Here "heating" is to be recognized as consideration of all thermal aspects of the problem.

Considering the factors shown above and the heterogeneous, transient nature of the mechanical, kinetic, and thermal components, the most favorable time for chemical reaction is during the loading pulse, not during the post-shock thermal-cooling period.

Energy Localization

Prior studies of dynamic compaction of powders to achieve high density compacts have devoted effort to development of models of localization of mechanical energy on the surfaces of powders to explain observations of local melting. Unfortunately, the models that have been developed are too narrowly focused and do not realistically consider basic materials response aspects of shock-compression processes. The models fail to account for the substantial observations that show results demonstrating that melting is not the universal, dominant process.

The energy localization problem is summarized conceptually in Fig. 6.2. In the problem, the energy input to a local region is introduced by the mechanical deformation. Energy leaving the region is transported by thermal conduction. The balance between these two processes at a given time leads to the variation in local temperature. For this reason, the time scale of the competing thermal input and thermal loss processes is critical. In times of tens or hundreds of nanoseconds, thermal conductivity of metals will dominate the process, and if the mechanical energy cannot be deposited rapidly, little, if any, localized increase in temperature will result.

It should be recognized that the smallest increment of energy will result in infinite temperatures if it is localized in a vanishingly small volume of material. Indeed, one can easily predict temperatures that vary by an order of

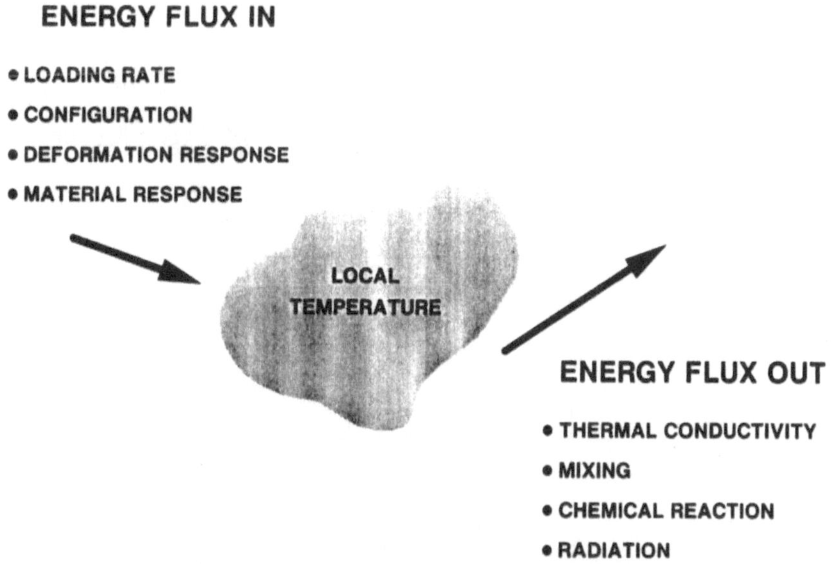

ENERGY FLUX IN

- LOADING RATE
- CONFIGURATION
- DEFORMATION RESPONSE
- MATERIAL RESPONSE

LOCAL
TEMPERATURE

ENERGY FLUX OUT

- THERMAL CONDUCTIVITY
- MIXING
- CHEMICAL REACTION
- RADIATION

Fig. 6.2. The energy localization problem in shock-loaded porous powder mixtures involves a balance between the rate at which energy is applied locally and the rate at which it is removed from the site.

magnitude by different assumptions as to the volume of material considered and the time of deposition of the energy.

It is well known, even in solid density samples, that effective shock viscosities of solids cause rise times to peak pressure that can be hundreds of nanoseconds; rise time is a materials and strain-rate dependent factor. In porous solids the rise times are expected to be even longer due to the propagation through the heterogeneous mixture of solids and voids. The influence of shock viscosity on energy localization is well described in various analytical work [87A02, 81K01, 81K02], which has apparently not been studied by investigators in dynamic compaction processes [85K01, 86G04, 84S01]. Realistic viscoplastic effects in "pore collapse" have been well described mathematically by Attetkov and Solov'ev [87A02]. Incorporation of realistic, materials property dependent processes into thermomechanical descriptions is essential to the development of correct physical models of energy localization in powder compacts. In the case of porous solid compacts it is clear that "temperature" is an ill defined word. Cited temperatures are strongly model dependent, and should be so indicated.

6.3 Mathematical Models

Horie and his coworkers [90K01] have developed a simplified mathematical model that is useful for study of the heterogeneous nature of powder mixtures. The model considers a heterogeneous mixture of voids, inert species, and reactant species in pressure equilibrium, but not in thermal equilibrium. The concept of the Horie VIR model is shown in Fig. 6.3. As shown in the figure, the temperatures in the inert and reactive species are permitted to be different and heat flow can occur from the reactive (usually hot) species to the inert species. When chemical reaction occurs the inert species acts to ther-

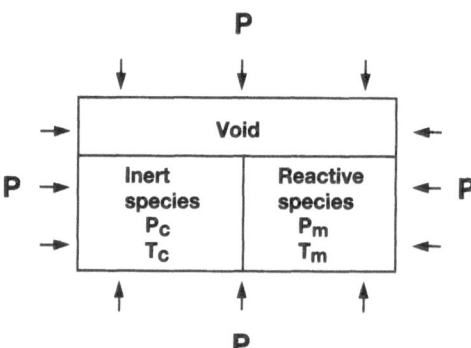

Fig. 6.3. The VIR Model of Horie considers the chemical reaction process in terms of three components: void, inert, and reactants. The influence of the inerts is critical as that component causes thermal quenching of incipient reactions.

mally quench the reaction. Pressures achieved are substantially influenced by the heat of reaction of the reactive species, as well as by the bulk modulus and thermal expansion of the inert species. As the local temperature is controlled by the rate at which thermochemical energy is released and the heat is transported into the inert species, strongly time dependent pressure pulses are predicted by model calculations. This model appears to hold promise in developing an understanding of the early processes occurring in shock-induced chemical reactions.

6.4 Shock Compression of Porous Powder Compacts

In most cases of interest, shock-induced chemical reactions in solids are studied in mixtures of powders of the potential reactants. In the earlier description of conceptual models it was emphasized that the pores provide space in which the potential reactants can be more intimately mixed in order

Table 6.2. Zero-pressure, shock-velocity intercepts of porous powder compacts.*

	$U = U_0 + au$			
Material	Compact $(g\,cm^{-3})$	Density (%)	U_0 (km s)	Reference
Al	2.01	50	0.4	Bakanova et al. [73B01]
Al	0.90	33	0.2	Bakanova et al. [73B01]
Cu	6.33	71	0.57	Trunin et al. [89T02]
Cu	4.47	50	0.15	Trunin et al. [89T02]
Cu	3.57	40	0.20	Trunin et al. [89T02]
Cu	4.67	52	0.2	Bakanova et al. [73B01]
Cu	3.00	33	~ 0.0	Bakanova et al. [73B01]
Cu	7.41	83	0.66	Boade [70B02]
Cu	6.05	68	0.51	Boade [70B02]
Fe		89	1.2	Herrmann [69H02]
Fe		74	0.8	Herrmann [69H02]
Fe		61	0.5	Herrmann [69H02]
Fe		42	0.2	Herrmann [69H02]
Fe		33	~ 0	Herrmann [69H02]
Fe		17	~ 0	Herrmann [69H02]
Ni	6.28	71	0.42	Trunin et al. [89T02]
Ni	4.43	50	0.46	Trunin et al. [89T02]
Mo	5.59	55	0.50	Trunin et al. [89T02]
Mo	4.67	52	0.50	Bakanova et al. [73B01]
W	12.6	68	0.53	Boade [70B02]
W	10.6	55	0.60	Bakanova et al. [73B01]
W	5.4	28	0.40	Bakanova et al. [73B01]
TiO_2	1.2	28	0.9	Pashov et al. [79P03]

* Note: Data are selected from studies with significant measurements at pressures less than about 10 GPa.

to become reactants in subsequent chemical reactions. In most cases, the densities of the mixed powder compacts are from 35% to 65% of solid density. Materials are expected to behave distinctively differently in such configurations due to the different mechanical and physical properties changed from the large deformation.

Consider, for example, the deformation history of the powder particles. To the extent that the powders deform plastically, the material must be deformed to strains of about 100%. This strain will be achieved in times significantly longer than at solid density with strain rates of perhaps 10^4. Thus, the deformation history is radically different from the same material at solid density. In this very large strain, lower strain rate regime, constitutive data to guide model calculations are limited. In the case of mixed powders, we expect the deformation histories to be different for each component, and in every case, the deformation history will be dependent on the particle morphology (size and shape). These situations are unlike that of a solid with pores (so-called "pore collapse" [87A02]) which can change local deformation in the vicinity of the pore, but is expected to have little effect on deformation in the bulk.

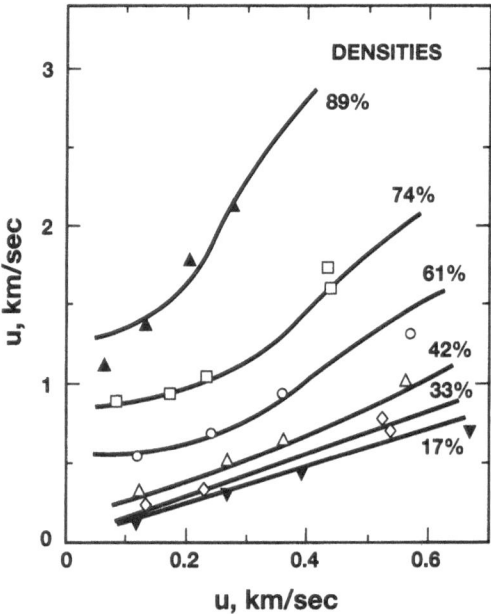

Fig. 6.4. At relatively low pressures the shock speeds observed for stress waves in low density powder compacts are dominated by the crush-up of the powder toward solid density. The figure shows measured wavespeeds for a range of densities and fits to the data based on Herrmann's *P-α* model on Fe. Note the unusually low wavespeeds compared to solid density (after Herrmann [69H02]).

In addition to these micromechanical considerations, low pressure shock compression of porous powder compacts has distinctive features not encountered in low pressure solid density samples. Basically, the sample is dominated by the pores, and the wavespeed at pressures less than those required to "crush" the sample to solid density is unusually low and is little dependent on the properties of the solid.

Table 6.2 summarizes the low pressure intercept of observed shock-velocity versus particle-velocity relations for a number of powder samples as a function of initial relative density. The characteristic response of an unusually low wavespeed is universally observed, and is in agreement with considerations of Herrmann's P-α model [69H02] for compression of porous solids. Fits to data of porous iron are shown in Fig. 6.4. The first order features of wavespeed are controlled by density, not material. This material-independent, density-dependent behavior is an extremely important feature of highly porous materials.

6.5 Sample Preservation Technique

In an experiment in which a sample is subjected to controlled shock loading and preserved for post-shock analysis, the shock-recovery experiment, the quantification, and the credibility of the experiment rest directly upon the apparatus in which the experiments are carried out. Quantification must be established with two-dimensional numerical simulation and this can only be accomplished if the recovery fixtures are standardized. The standardized fixtures must be capable of precise assembly so that the conditions actually achieved in the experiment are those of the simulation.

The author's work has included the development of the Sandia Bear and Bertha explosive recovery fixtures, that provide a standardized set of fixtures in which recovery experiments can be routinely carried out at peak shock pressures from 4 to 500 GPa. Shock-induced, mean-bulk temperatures from 50 to 1200 °C are achieved with variation in the density of the powder compacts under study.

Use of the term "mean-bulk temperature" is to define the model from which temperatures are computed. In shock-compression modeling, especially in porous solids, temperatures computed are model dependent and are without definition unless specification of assumptions used in the calculations is given. The term mean-bulk temperature describes a model calculation in which the compressional energy is uniformly distributed throughout the sample without an attempt to specify local effects. In the energy localization case, it is well known that the computed temperatures can vary by an order of magnitude depending on the assumptions used in the calculation.

A schematic drawing of the system is shown in Fig. 6.5. A powder sample is pressed in place in a cavity in a copper capsule. The cavity is closed with

HIGH EXPLOSIVE LOADING

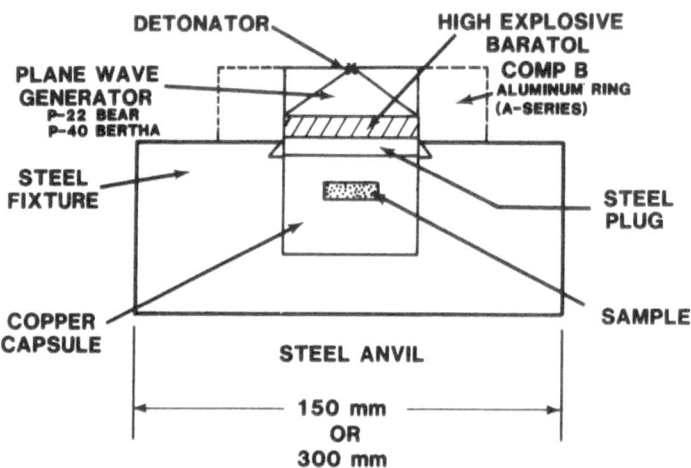

Fig. 6.5. The shock-compression conditions imposed on powder compacts preserved for post-shock analysis are controlled by details of the shock-recovery fixtures. In all the work of Chap. 6, the Sandia "Bear" and "Bertha" fixtures are used. The fixtures represent a standardized system that is highly reproducible and has been subject to extensive numerical simulation.

removable plugs in both ends. Both surfaces of the plugs are mechanically "lapped" to optical flatnesses prior to assembly. Pressing the powder in place assures that there are no voids or gaps within the system. After the plugs are in place, both surfaces of the capsule are lapped to optical flatnesses.

The sample capsule is placed in a tight-fitting 4340 steel fixture that serves to support the copper capsule. Pressure from the detonation of the explosive is transmitted to the copper capsule through a mild steel driver plate. This plate is also lapped optically flat on both surfaces. The mild steel acts to shape the pressure pulse due to the 13 GPa structural phase transition. With proper choice of the diameter of the driver plate and beveled interior opening of the steel fixture, shock deformation of the driver plate acts to seal the capsule within the fixture.

The high explosives, baratol or Composition B-3, are used to produce the plane wave loading into the driver plates. These explosives have been widely studied in substantial work at Los Alamos. Plane waves are introduced into the explosive pads with either P-22 or P-40 plane-wave generators developed at Los Alamos. The Bear system is based on the 56 mm diameter of the P-22, while the larger sample size Bertha system is based on the 102 mm diam of the P-40. More details on sample dimensions are reported by Graham [87G03].

It should be observed that every element except the powder system in the recovery system is chosen for favorable shock properties which can be confidently simulated numerically. The precise sample assembly procedures assure that the conditions calculated in the numerical simulations are actually achieved in the experiments. The influence of various powder compacts in influencing the shock pressure and mean-bulk temperature must be determined in computer experiments in which various material descriptions are used. Fortunately, the large porosity (densities from 35% to 75% of solid density) leads to a great simplification in that the various porous samples respond in the same manner due to the radial loading introduced from the porous inclusion in the copper capsule.

Numerical Simulation

One-Dimensional Numerical Simulation

Although two-dimensional features are readily apparent in the shocked and preserved samples, and study of the loading conditions indicates the need for two-dimensional studies, the influence of various powder descriptions and time scales of the events can be studied more efficiently in one-dimensional

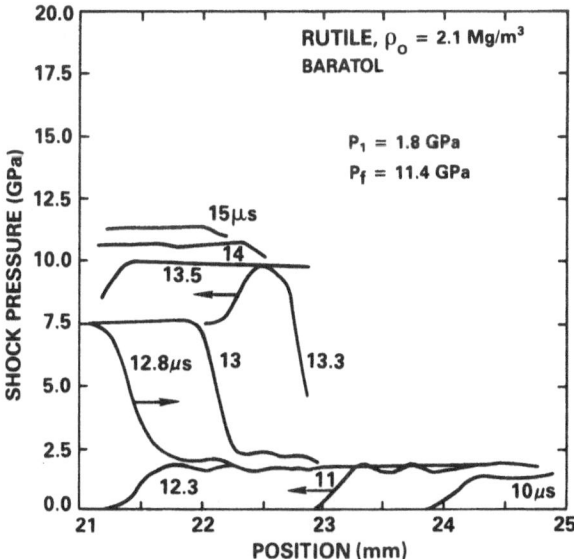

Fig. 6.6. One-dimensional, pressure-versus-location predictions at various times are shown for a typical powder compact subjected to baratol plane-wave explosive loading. The pressure is shown to "ring-up" to a final value between the copper end plates (after Graham [87G03]).

simulations, which serve as computer experiments. These simulations have been carried out in the Sandia Lagrangian computer code CHARTD [87G03].

For the two explosive loading systems used, the initial pressure wave into the powder is relatively low, varying from perhaps 1.5–4 GPa. In such cases the most relevant compression characteristic of the powder compact is its "crush strength", i.e., the pressure required to compress the porous compact to solid density. In the simulations, this strength can be varied over a wide range with the P-α model. The wavespeed of the initial waves was modeled on the basis of shock-compression data on rutile at densities from 44% to 61% of solid density [74T02].

Typical pressure and temperature histories computed are shown in Figs. 6.6 and 6.7. In Figs. 6.6, the pressure is shown as a function of position within the powder compact at various times. For the baratol explosive loading shown, an initial wave, whose pressure is 1.8 GPa, is shown moving slowly from right to left. Upon reflection from the rear interface with the copper, the pressure jumps to a much higher value and then quickly reverberates to a peak pressure of about 11.4 GPa. The shorter reverberation time reflects the higher wavespeed and the major reduction in thickness in the compressed powder.

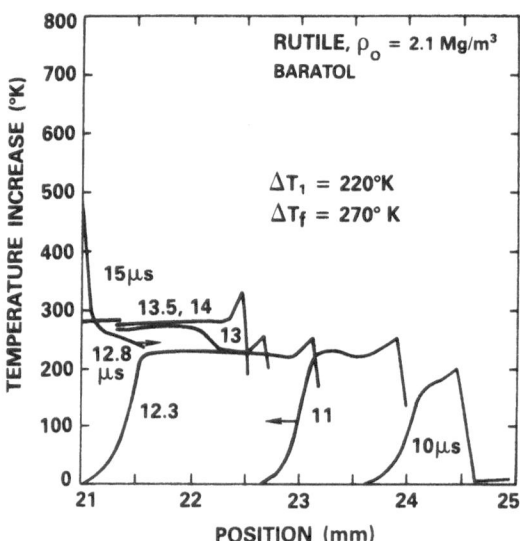

Fig. 6.7. The predicted, one-dimensional, mean-bulk temperatures versus location at various times are shown for a typical powder compact subjected to the same loading as in Fig. 6.5. It should be observed that the early, low pressure causes the largest increase in temperature due to the "crush-up" of the powder to densities approaching solid density. The "spike" in the temperature shown on the profiles at the interfaces of the powder and copper is an artifact due to numerical instabilities (after Graham [87G03]).

As shown in Fig. 6.7, typical temperature histories show a quite different behavior from that observed with pressure in that the initial low pressure wave produces the major portion of the increase in temperature. This behavior is the expected consequence of the large volume compression of the powder compact.

As shown in Fig. 6.6, the peak pressures are little affected by the density of the powder. This behavior is characteristic of a reverberation process to achieve peak pressure. On the other hand, the mean bulk temperature is strongly affected by the powder density, representing the volume compression to achieve solid density.

The influence of crush strength on mean-bulk temperature achieved in one-dimensional compression of powders has been extensively investigated

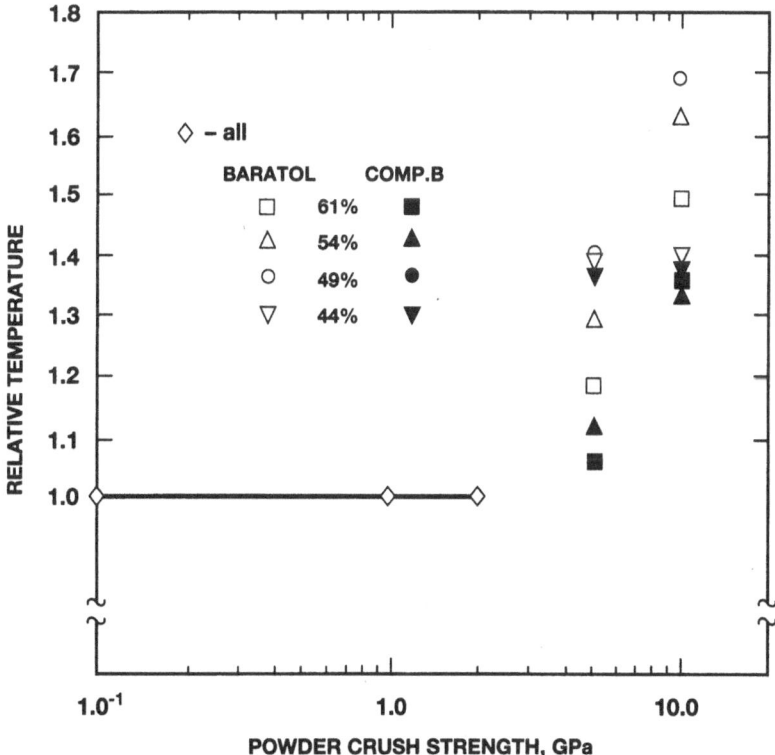

Fig. 6.8. The peak mean-bulk temperatures predicted in one-dimensional numerical simulation are investigated for powder compacts of different "crush strengths." For the explosive loadings of the Bear fixtures, no difference in temperature is predicted for crush strengths up to about 2 GPa. This value is about that of the initial loading wave into the samples. Above that pressure the crush strength has a strong effect on temperature. The predicted behavior can be understood in terms of the various loading paths.

in the one-dimensional simulations. In these simulations the crush strength of the powder was varied from 0.1 to 20 GPa at powder densities of 61%, 54%, 51%, and 44% of solid density. Peak temperatures were calculated for loading with baratol and Composition B. These computer experiments with results as shown in Fig. 6.8, show for crush strengths up to 2 GPa that there is no effect of crush strength on peak temperature. This can be explained from recognition that at input pressures greater than the crush strength the volume compression achieved by the shock loading is the same. Such a situation will result in mean-bulk temperatures that are unaffected by crush strength.

At crush strengths of 5 and 10 GPa the temperatures are larger than those for lower crush strengths. Again, the observed temperatures can be explained by the lack of compaction for the initial loading wave and a larger area under the pressure-volume curve as the loading proceeds along the incompletely

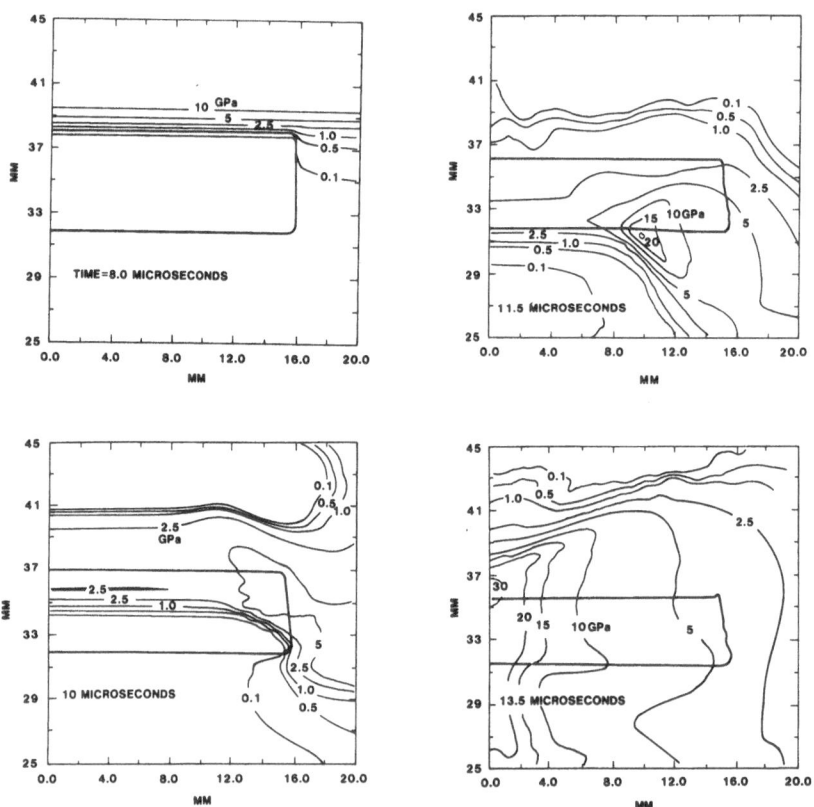

Fig. 6.9. Two-dimensional numerical simulations are depicted for the Sandia Momma Bear fixture. Pressure contours within one-half the powder compact are shown at various times. The principal feature shown is the development of a radial-mode loading due to the low shock impedance of the powder (after Graham [87G03]).

compressed powder. The critical feature affecting the temperature is whether the crush strength is less than the initial loading pressure. Other thermodynamic properties of the powder materials have a secondary effect.

Two-Dimensional Numerical Simulation

Realistic description of the conditions that the powder compacts experience must be based on two-dimensional numerical simulations, which have been carried out on the Sandia fixtures [84G01, 86G04]. The background established from the one-dimensional simulations allows the work to be carried out more efficiently. In this book, the simulations were carried out with the Sandia two-dimensional Eulerian computer code, CSQ.

Because of its emphasis on consistent thermodynamics, the CSQ code does not permit the use of a P-α model for the crush-up behavior of the powder. Thus, it was necessary to draw upon the experience in the one-dimensional simulation to select appropriate shock-compression materials behaviors. The

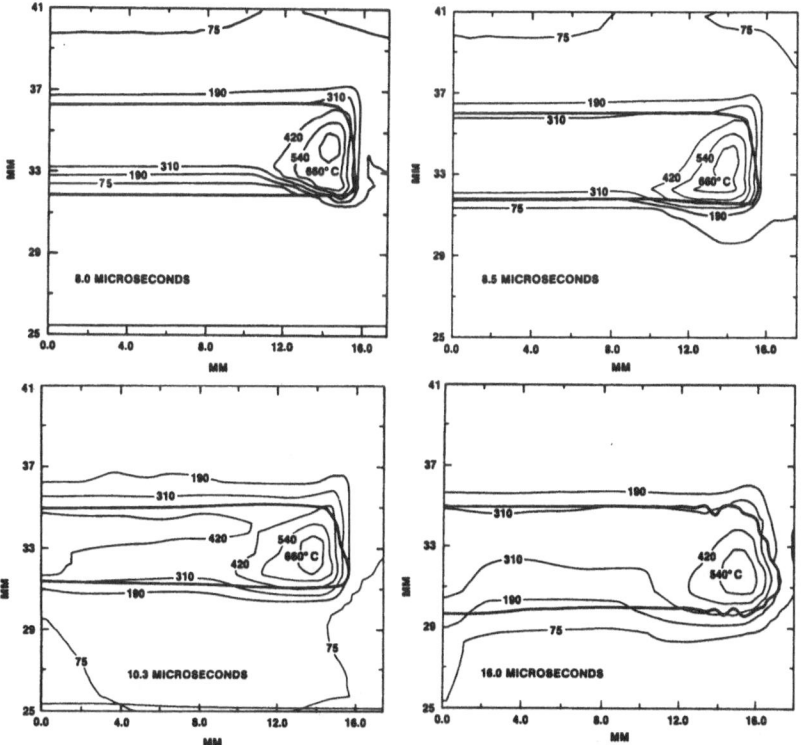

Fig. 6.10. Two-dimensional simulations of the same arrangement of Fig. 6.9 show the predicted mean-bulk temperature contours at various times. The principal unusual feature is the hot region developed in the outside area due to simultaneous longitudinal and radial loading (after Graham [87G03]).

simulations chose a "rutile" powder ceramic material based on its solid density and parameters for initial shock velocity U. Beyond the use of unusually low shock-velocity values, no further attempt was made to treat the sample as a porous material. Thermodynamic parameters were selected to phenomenologically match data available for shock compression of rutile powder. For certain simulations, alternate descriptions were used to confirm the one-dimensional observations that the peak pressures are not strongly dependent on the powder description. Mean bulk temperatures can be made to vary in an absolute sense by various powder descriptions by perhaps 20%, but the relative temperatures of a particular simulation remain the same with the various powder descriptions.

Pressure contours for a simulation of a Momma Bear fixture loaded with baratol are shown in Fig. 6.9. The figure shows pressure contours at four different times, emphasizing how the initially planar loading is converted to a radial loading by the high wavespeed in the copper compared to the powder. In Fig. 6.10, a similar set of temperature contours is shown for the same situation. In this case the dominant feature is the region of elevated temperature in the outer region of the sample. This feature is prominent in examination of shocked and recovered samples; indeed, all prominent features revealed in the simulations are found to be present in the samples. Considerable confidence has been developed in the simulations with the experience of examining hundreds of powder samples. Excellent reproducibility

Table 6.3. Sandia Bear and Bertha fixture characteristics.

Fixture	Peak pressure (GPa)		Mean-bulk temperature (°C)	
	Bulk	Focus	Bulk	Edge
Baratol explosive				
Baby Bear	20 ± 6	40	250	310
Momma Bear A	16 ± 4	32	225	310
Momma Bear	7.5 ± 2.5	27	225	250
Poppa Bear	5 ± 1	4.5	150	75
Bertha A	7.5 ± 2.5	18	200	200
Big Bertha A	6 ± 1.5	11	200	200
Bertha	4 ± 0.5	4.6	75	75
Composition B explosive				
Baby Bear	27 ± 5	45	225	950
Momma Bear A	22 ± 4	46	400	650
Momma Bear	18 ± 3	41	425	550
Poppa Bear	8 ± 1.5	9	375	300
Bertha A	2.3 ± 3		⋯	⋯
Big Bertha A	13 ± 3	19	425	550

Mean-bulk temperatures are representative of a powder compact at a density of 55% of solid density.

is also indicated by the consistency of material characteristics in various samples prepared by the same fixture and loading at various times.

A summary of peak pressure and mean bulk temperatures in the various fixtures is shown in Table 6.3. Included in the characterization is the peak pressure along the axial few millimeter region along the axis of the samples (called focus) for which the radial focusing produces a high pressure region for a period of about 100 ns.

CHAPTER 7

Shock Modification and Shock Activation: Enhanced Solid State Reactivity

In this chapter: shock modification of powders (their specific area, x-ray diffraction lines, and point defects); measurements via analytical electron microscopy, magnetization and Mössbauer spectroscopy; shock activation of catalysis, solution, solid-state chemical reactions, sintering, and structural transformations; enhanced solid-state reactivity.

7.1 Shock Modification

If the processes controlling solid state chemistry are to be identified and quantified, the properties of the solids involved as reactants must be described. Because solid state chemistry is largely controlled by the morphology and defect state of the reactants, the influence of shock compression on these factors must be quantified. Many of the reactants of interest are inorganic refractories for which there have been little data on shock modification; even for metals, most of the data on shock modification are for large, confined solid density samples. In most of the chemical studies, the reactants are in the form of powder particles in low density compacts. In powder form, deformation and plastic flow differs significantly from that in bulk form. Thus, studies of shock-modified powders are required to develop an understanding of the defect state of reactants. Results of changes in specific surface area investigate the degree of powder comminution. Studies of x-ray diffraction line broadening provide a comprehensive description of lattice strain resulting from shock-induced defects. Studies of point defects in selected substances allow comparison of the shock-induced defects with those encountered in more conventional defect studies. Magnetization provides a sensitive measure of disorder induced by shock deformation, while Mössbauer spectroscopy gives a direct measure of local atomic states.

The reported shock-modification observations show that shock-treated powders are substantially modified in their defect structures. From the defect point of view they are essentially new materials. Concentrations of point, line, and higher-order defects are found to be as large as those achieved by any process. In addition, effects such as reduction in crystallite size create materials in the mesoscopic size range in which surface energies are significant. The observed modifications lead to the conclusion that solid state chemical activity should be substantially enhanced. Nevertheless, the defect concentrations and morphologies are so far from those reported in other work that there is little independent basis to directly relate the defects to estimates of resulting solid state reactivity. To complete the descriptions of potential reactants it is necessary to carry out studies of enhanced solid state reactivity on shock-modified powders. With measures of shock modification and solid state reactivity in hand, chemical synthesis between reactant powders can then be studied to develop the full picture of shock-induced solid state chemistry.

The measures of solid state reactivity to be described include experiments on solid–gas, solid–liquid, and solid–solid chemical reaction, solid–solid structural transitions, and hot pressing–sintering in the solid state. These conditions are achieved in catalytic activity measurements of rutile and zinc oxide, in studies of the dissolution of silicon nitride and rutile, the reaction of lead oxide and zirconia to form lead zirconate, the monoclinic to tetragonal transformation in zirconia, the theta-to-alpha transformation in alumina, and the hot pressing of aluminum nitride and aluminum oxide.

Specific Surface Area

One of the early proposals for greatly enhanced reaction rates in shock-compressed powders was the efficient comminution of the powders leading to substantially increased surface area of the reactants. Such a process appeared particularly appealing in the inorganic refractory materials whose deformations most often are brittle and subject to fragmentation. As powders are typically complex in morphology, changes in specific surface area are best investigated with the BET specific surface method involving adsorption of nitrogen. Williams and his co-workers [85L01, 86W04] have studied specific surface areas of a number of shock-modified powders.

Results of investigations of shock-induced specific surface changes are summarized in Table 7.1. In the table, the data are summarized in terms of the maximum value of specific area observed and the pressure at which the maximum is observed. The specific surface at the highest shock pressure is also indicated.

What is observed is that there are significant changes in specific surface, but that they are relatively modest and cannot account for large changes in reaction rates in shocked powders. The observed behavior can be characterized into typical behaviors as summarized in Fig. 7.1. If comminution is the dominant behavior, the specific surface area will be observed to increase. Such a behavior is called "Type a." If consolidation is the dominant behavior, specific surface area will be observed to decrease. Such a behavior is called "Type b." In the most typical case, the specific surface increases at low pres-

Table 7.1. Shock-modified powders: Specific surface areas.

Substance	Specific surface area $(m^2 g^{-1})$		
	Starting	Maximum (pressure)	Highest (pressure)
AlN	2.2	2.8 (17)	1.0 (27)
Si_3N_4	4.1	6.7 (20)	6.1 (27)
TiC	2.0	3.5 (17)	3.4 (27)
TiB_2	0.6	1.8 (22)	1.8 (22)
Al_2O_3	8.1	9.7 (17)	6.9 (27)
$Al_2O_3^{-2}$	...	1.3 (17)	0.6 (27)
MnO_2	0.94	4.0 (20)	2.3 (27)
Fe_3O_4	10.9	10.9 (0)	1.3 (27)
Fe_2O_3	5.4	6.9 (17)	2.7 (22)
ZnO	6.0	6.6 (0)	1.6 (22)
ZrO_2	8.4	7.8 (20)	7.8 (20)
TiO_2 mixed	2.4	4.2 (20)	...
Rutile-1	1.8	2.5 (20)	1.7 (27)
Rutile-2	0.3	1.3 (20)	1.0 (27)

Type a: Comminution; TiB_2. Type b: consolidation; ZrO_2, ZnO, Fe_3O_4.
Type c: Comminution consolidation; AlN, TiC, Al_2O_3, TiO_2, Si_3N_4, Fe_2O_3, MnO_2.
Pressures are in GPa.

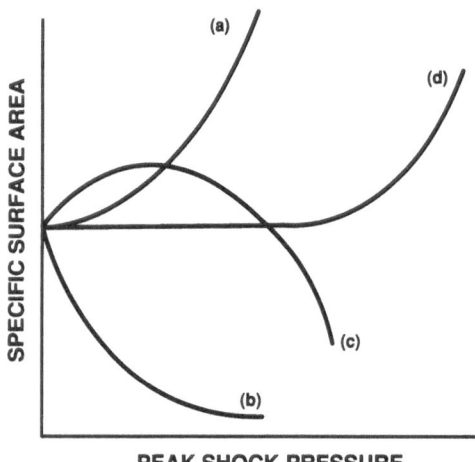

Fig. 7.1. Observed specific surfaces for shock-modified powders show four typical behaviors indicative of (a) comminution, (b) consolidation bonding, (c) comminution followed by bonding, and (d) comminution after phase transformation [85L01].

sure, indicating comminution. At higher pressure, there is a subsequent decrease in surface area, indicating high pressure bonding. Such a behavior is described as "Type c" in the figure.

Because of the large number of powders studied and their generally refractory character, it appears that it can safely be concluded that substantial particle comminution does not occur in shock-compressed powders. Rather, the process is better characterized by changes in morphology and limited interparticle bonding.

X-Ray Diffraction Line Broadening

Morosin and his co-workers [84M01, 87M01] have studied x-ray diffraction on a number of shock-modified powders in considerable detail. It should be recognized that the technical problem of line broadening in shock-modified powders is more difficult than in other situations in that both crystallite size and strain are substantial and both influence broadening. Furthermore, the materials studied are typically of lower symmetry than metals in which such studies are more typically carried out. The lower symmetry also results in significant anisotropy in the residual strain, further complicating the analysis. Given these complications, generally accepted procedures and assumptions are suspect and require careful examination, including the development of new analytical procedures.

Table 7.2 summarizes the principal x-ray diffraction line broadening observations. The most detailed studies are on TiC, which is especially interesting because of its high symmetry. Other interesting materials studied in some detail include Al_2O_3, TiO_2, LaB_6, and ZrO_2. Figure 7.2 summarizes the observed residual strain induced by a given shock pressure for various substances.

What is especially notable in all the materials is that the residual strain values are those typically observed in cold-worked metals. Even though the investigated materials are generally considered brittle, the confined rapid deformation conditions of the present work result in a large plastic deformation whose saturation plastic deformation values appear to be the largest achievable in the materials. Thus, the shock-loading has produced a highly defective state with large energies stored in residual strain. Perhaps as significant, the strain is significantly anisotropic.

Table 7.2. Shock-modified powders: X-ray diffraction line broadening.

Substance	Saturation strain 10^{-3} (pressure)	Crystallite size	
		initial	minimum (pressure)
TiC	1.5 (27)	700	300 (27)
LaB_6	1.7 (22)	610	210 (22)
TiB_2	3 (27)	...	...
AlN	3 (27)	...	...
Al_2O_3	5 (27)	...	300 (27)
TiO_2	3 (17)	...	560 (27)
Fe_2O_3	1.4 (17)	~200	56 (17)
ZrO_2	4 (20)	600	190 (27)

Pressures are in GPa. Crystallite sizes are in angstroms.

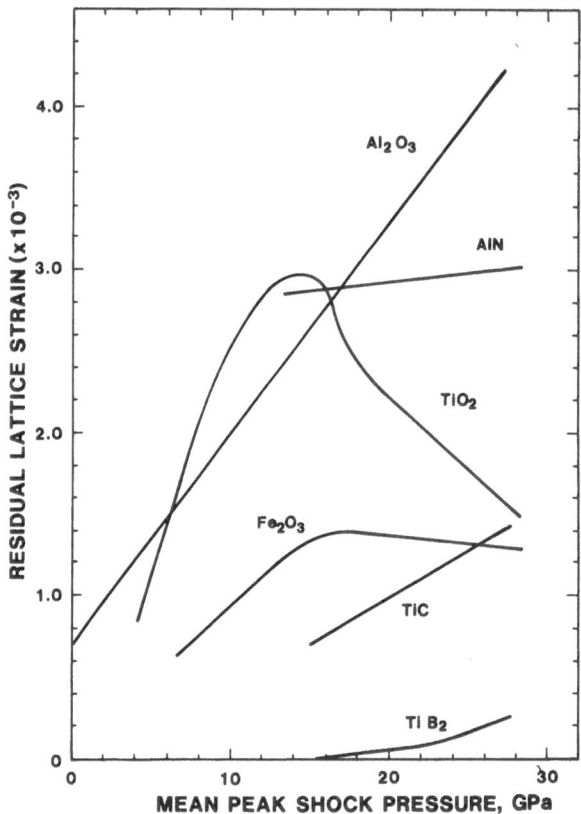

Fig. 7.2. X-ray diffraction line broadening studies in inorganic powders by Morosin and co-workers show evidence for large plastic deformation with residual strain characteristic of cold-worked metals [86M02].

The large plastic deformation acts to reduce the coherent scattering area—crystallite size—values as shown in Fig. 7.3. At the highest pressure conditions, crystallite sizes are typically reduced to a few hundreds of angstroms. The crystallite sizes are anisotropic. Considerable surface energy is associated with sizes in this range and would be expected to have a strong effect on chemical reactivity. Such sizes have been described as "mesoscopic" to indicate they are larger than typical atomic dimensions and less then macroscopic sizes. Mesoscopic crystallite size materials have been observed to exhibit significant surface-area effects.

In a comparison of shock-modified powder to powder subjected to other intense deformation, data on shock-modified TiC was compared to a well annealed TiC powder wet milled for many hours to similar values of residual strain. As depicted in Fig. 7.4 the anisotropies observed in residual strain and crystallite size in the two cases are quite different. The shock-modified powders show less anisotropy in strain and more anisotropy in crystallite size

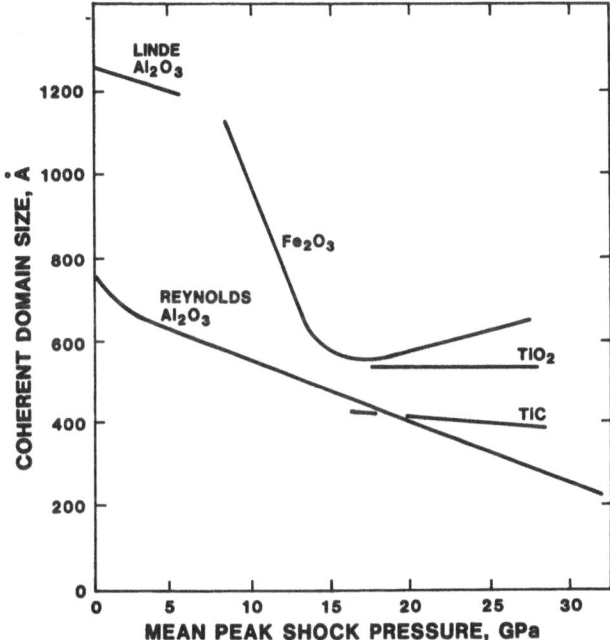

Fig. 7.3. Crystallite size determined from x-ray diffraction line broadening studies show substantial shock-induced reductions. The chemical reactivity of such powders would be expected to be greatly enhanced [86M02].

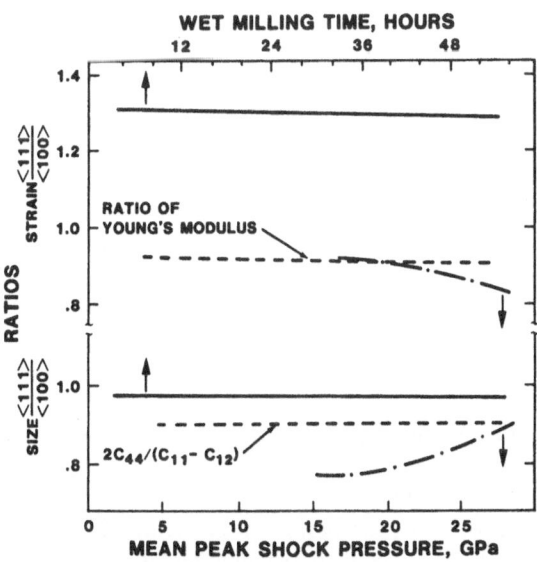

Fig. 7.4. Residual strain and crystallite size are compared for TiC powders subjected to wet milling and shock modification. Significant differences are observed in the anisotropies of both features (after Morosin and co-workers [86M02]).

than the milled powders. This observation shows clear evidence for differences between shock-modified and cold-worked powders. It should also be observed that anisotropy in residual strain and crystallite strain are larger in alumina and rutile as expected from the lower symmetry of the oxides.

In a similar study on shock-modified and jet-milled LaB_6, [89Z01], it was found that shock processing had a larger influence on microstructure than jet milling. Comparison of the jet-milled and 5 GPa shock-modified powder showed strong similarities. At higher shock pressures, however, crystallite sizes were greater and strain values smaller for the jet-milled materials.

Point Defects—Electron Spin Resonance

Evidence for large concentrations of point defects has been widely reported in studies of shock-modified solid density metals. The large concentrations are thought to result from rapid deformation, which has been predicted to lead to efficient production of point defects. Little information has been available on inorganic materials, but recent work on shock-modified, high-purity rutile by Venturini and co-workers [84V01] in both powder and single crystal form has shown clear evidence for unusually high concentrations of point defects and their resulting changes in physical properties.

Rutile is an ideal material for study of point defects. It has been extensively studied by electron spin resonance (ESR), both as a host lattice for paramagnetic impurity ions and as a defect structure. The starting material of the present work was found to contain no resonances, and was white in color. The shock-modified powder was found to be dark grey in color to a degree depending upon the shock conditions, and powder in the higher-temperature, outer-edge region is typically noticeably darker in color. Existence of color centers is well documented in this material.

Shock-modified rutile is found to exhibit two characteristic resonances, which can be confidently identified as (1) an isotropic resonance characteristic of an electron trapped at a vacancy, and (2) an isotropic resonance characteristic of a Ti^{+3} interstitial. The data indicate a concentration of $2 \times 10^{19} \, cm^{-3}$, which is an order of magnitude greater than observed in hydrogen- or vacuum-induced defect studies. At higher pressures the concentration of interstitials is the same as at lower pressure, but more dispersion is observed in the wave shape, indicating higher microwave conductivity.

Figure 7.5 shows measured concentrations of the interstitials in reduced rutile from previous studies by Hasiguti [72H01]. It has been the consistent observation in prior work that, above some level of concentration, both concentrations and conductivity are observed to reverse their lower defect level behavior. Thus, this wide band-gap material will not form sufficiently high concentrations of defects to become metallic as in germanium and silicon. It is thought that the higher concentration of defects results in formation of higher-order, oxygen-defect complexes characteristic of crystallogra-

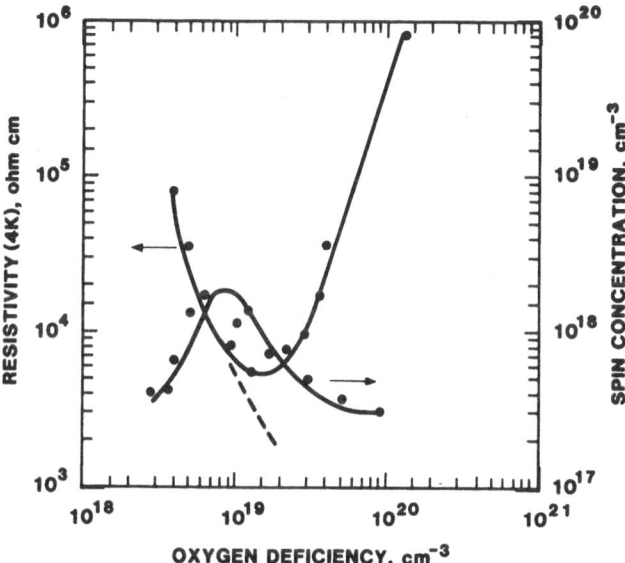

Fig. 7.5. Previous studies summarized by Hasiguti [72H01] on oxygen deficient TiO_2 treated to achieve a wide range of oxygen deficiencies show large changes in low temperature resistivity and Ti^{3+} interstitial concentrations. Both reach minima or maxima at concentrations of about 10^{19}. The change in behavior is thought to be due to formation of large oxygen-defect complexes. Studies of shock-modified TiO_2 by Venturini and co-workers show spin concentrations of 2×10^{19} cm^{13}, about an order of magnitude larger than in prior studies [84V01].

phic shear. In the present case, an order-of-magnitude higher concentration of defects is formed.

Further evidence for the unique nature of the shock-formed point defects is the dispersion in ESR lineshape characteristic of conductivity at temperatures above 30 K. In shock-modified powder the conductivity is constant down to 2 K, indicating that the electrons responsible for the conductivity are not trapped. These observations indicate that shock-modified rutile is in a physical defect state that has not been obtained in more conventional vacuum-reduction defect studies.

Analytical Electron Microscopy

Carr and his co-workers [86C01, 87C01] have shown that transmission electron microscopy is a powerful tool in characterizing linear and higher-order defect configurations and their densities on shock-modified rutile, alumina, aluminum nitride, and zirconia [84H02]. The principal impediment to detailed characterization of shock-formed defects is their very high concentrations, which prevent identification of specific deformation features except in special cases. The microstructure of shock-modified rutile powder was found to show a variety of defects with their density and nature dependent on the shock conditions. Only at the lowest pressure of 5 GPa was it possible to image individual features. At higher pressures, the density of defects was so great that individual features overlapped. It was found that the microstructure of the shock-deformed rutile was complex and contained a number of familiar defects, along with an unreported defect and the absence of crystallographic shear, which is perhaps the most characteristic feature of conventionally deformed rutile.

Evidence was found for both twinning and slip within a twinned grain, indicating that the deformation mode under shock loading is sensitive to crystallographic orientation. Slip traces observed parallel to {100} are unique because they show that new deformation modes may be activated under shock loading, resulting in defects that do not form easily under conventional loading. It may be that under conventional loading, any slip occurring on {100} planes nucleates cleavage, while under the large component of hydrostatic pressure of the present work such a nucleation is prevented.

In one of the most significant observations, small amounts of recrystallized material were observed in rutile at shock pressure of 16 GPa and 500 °C. Earlier studies in which shock-modified rutile were annealed showed that recovery was preferred to recrystallization. Such recrystallization is characteristic of heavily deformed ceramics. There has been speculation that, as the dislocation density increases, amorphous materials would be produced by shock deformation. Apparently, the behavior actually observed is that of recrystallization; there is no evidence in any of the work for the formation of amorphous materials due to shock modification. Similar recrystallization behavior has also been observed in shock-modified zirconia.

Although only limited microstructural features can be identified in shock-modified powders with transmission electron microscopy, the observations are in general agreement with the x-ray diffraction observations that indicate unusually dense concentrations of defects. Some evidence for unique features is reported. In none of the cases examined is there any evidence for localized melting at particle interfaces or significant interparticle bonding. There is significant evidence for heterogeneities in deformation mode within particular particles.

Magnetization and Mössbauer Spectroscopy

Hematite (α-Fe_2O_3) is an antiferromagnetic material exhibiting weak ferromagnetism above the Morin temperature (260 K), where it undergoes a magnetic spin-flip transition. The weak ferromagnetism effect is due to spin canting due to in-plane anisotropy. Because of a delicate balance of two competing mechanisms to magnetic anisotropy, the transition is very sensitive to material modifications.

A commercially available hematite powder has been subjected to con-

Table 7.3. Shock-modified powders: Crystallite size, strain, and static magnetization data on hematite (after Williamson et al. [86W03]).

Sample	Crystallite size (nm)	Strain (10^{-3})	M_0 (300 K) (emu g^{-1})	M_0 (7 K) (emu g^{-1})
As received	~ 200	<0.1	0.274	0.036
8 GPa	110	0.63	0.194	0.067
17 GPa	56	1.4	0.197	0.149
27 GPa	70	1.3	0.945	0.970

trolled shock loading at peak pressures between 8 and 27 GPa and studied with both static magnetization by Venturini and co-workers and Mössbauer spectroscopy measurements [86W02, 86W03] by Williamson and coworkers. Details of the Mössbauer spectra have been studied from 75 to 300 K, whereas the magnetization has been studied between 7 and 300 K. The use of both atomic level and macroscopic physical probes along with x-ray diffraction provides an unusually sound basis for interpreting the effects of shock modification.

X-ray diffraction measurements on the shock-modified samples show no evidence for new phases but show the typical shock-induced increase in residual stain and decrease in crystallite size. At room temperature, the magnetization measurements indicated decreased net magnetization as shown in Table 7.3 due to local shock-induced disorder. The low temperature magnetization shows a minimum at 17 GPa followed by a dramatic increase at the highest pressure of 27 GPa. The data for the 27 GPa sample indicate the formation of about 1% magnetite. The observed decrease in magnetization at 17 GPa is consistent with the sensitivity of magnetization to both residual strain and reduced crystallite size. At 7 K, the magnetization is observed to increase monotonically with shock severity to 17 GPa before the large increase at 27 GPa due to the formation of magnetite.

As shown in Fig. 7.6, the Mössbauer data show a reduction in Morin transition temperature with increasing shock severity. At temperatures below the transition, increasing shock severity causes greater retention of the higher temperature, weak ferromagnetic contribution. The measure of weak ferromagnetic (WF) fraction (the high temperature form) is a sensitive indication of shock modification.

The shock-modified hematite samples showed changes due to point, line, and higher-order defects as well as to their resulting macroscopic anisotropic strain fields, which can interact with local magnetic moments through magnetostriction. It was concluded that the following observed property changes were a direct consequence of shock-induced defects: (1) absence of a Morin transition for a large fraction of iron sites; (2) increased thermal hysteresis of the transition; and (3) incomplete spin reorientation for the antiferromagnetic state. The magnitude of the effects is similar in character and of the magni-

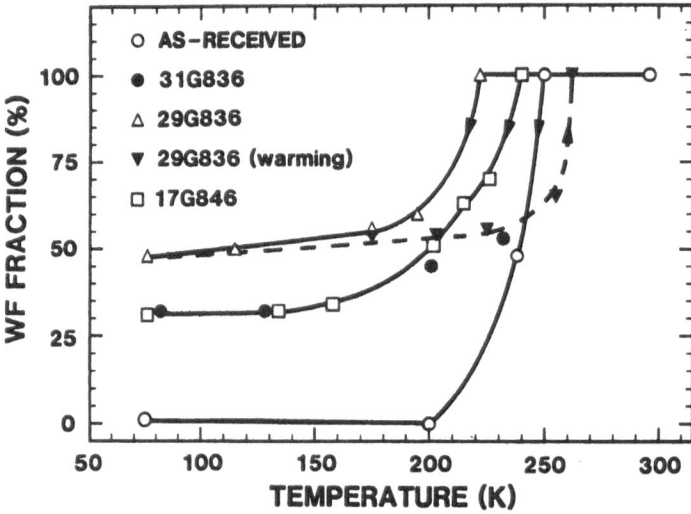

Fig. 7.6. The weak ferromagnetic (WF) fraction (high temperature form) of hematite provides a sensitive measure of shock modification. Sample 31G836 is an 8 GPa experiment. Sample 29G836 is a 17 GPa experiment, while 17G846 is a 27 GPa sample (after Williamson et al. [86W03]).

tude caused by the addition of about 5 mol % of alumina. The effects of reduced crystallite size are thought to be reduction of the Morin transition temperature, and increased magnitude of the room temperature quadrupole splitting. Observed increased hysteresis of the Morin transition is thought to be due to both crystallite size reduction and lattice strain. The retention of material that does not undergo the transition is perhaps best explained by local defects, while thermal hysteresis is best explained in terms of the macroscopic residual strain field.

The magnetization data, when combined with the Mössbauer data, show quantitative evidence for between 0 and 0.07 mol % of high magnetic moment defects at pressures less than 17 GPa. Large residual strain leads to an apparent decrease in net magnetization at high temperature due to strain-induced magnetic hardening.

Ferrites

A number of ferrites have been subjected to shock modification and studied with x-ray diffraction as well as static magnetization and Mössbauer spectroscopy [87V01]. Studies were carried out on cobalt, nickel, and copper ferrites as well as magnetite (iron ferrite).

As in other shock-modified powders, the x-ray diffraction measurements showed large values of residual strain resulting from extensive plastic deformation. At low magnetic field, the magnetization response was dominated by the magnetocrystalline, magnetostrictive, and shape anisotropy produced by the residual strain. At high magnetic fields, saturation magnetization was found to be significantly reduced. The magnetic hardness was increased for the copper, cobalt, and nickel ferrites and decreased for the iron ferrite.

A substantial portion ($\sim 15\%$) of the magnetite was found to be converted to hematite. As prior work showed small conversion (1%) of hematite to magnetite, the data indicate that the conversion can proceed in either direction depending upon the local microscopic deformation history of the powder particles.

Static magnetization and microwave loss has been studied in a high purity nickel ferrite and a barium ferrite [88V01]. The properties of both were found to be substantially modified. In the nickel ferrite, the shock-induced residual strain was found to cause a threefold increase in magnetic anisotropy through magnetostrictive coupling. This same effect also resulted in a large shift in the microwave loss peak at high frequency (6–12 GHz), a decreased loss below 9 GHz, and enhanced loss above 9 GHz.

The barium ferrite was found to have an increase in magnetic anisotropy, as in the nickel ferrite, but its overall effect on magnetization was less because of greater magnetocrystalline anisotropy. The shock modification caused reduced crystallite size and local damage that resulted in increased microwave absorption.

Shock-Modified Solids

The various studies of shock-modified powders provide clear indications of the principal characteristics of shock modification. The picture is one in which the powders have been extensively plastically deformed and defect levels are extraordinarily large. The extreme nature of the plastic deformation in these "brittle" materials is clearly evident in the optical microscopy of spherical alumina [85B01]. In these defect states their solid state reactivities would be expected to achieve values as large as possible in their particular morphologies; greatly enhanced solid state reactivity is to be expected.

It is particularly significant that no evidence is found for localized melting at particle interfaces in the inorganic materials studied. Apparently, effects commonly observed in dynamic compaction of low shock viscosity metals are not obtained in the less viscous materials of the present study. To successfully predict the occurrence of localized melting, it appears necessary to develop a more realistic physical model of energy localization in shock-compressed powders.

Based on the modification studies of this chapter, the potential reactants in the proposed model of important factors described in Fig. 6.1 must certainly be considered to be in a different configuration in terms of morphology and relative position than in the starting configuration.

7.2 Shock Activation: Enhanced Solid State Reactivity

Catalytic Activity

The catalytic properties of the shock-modified rutile whose defect properties have been reported in previous sections of this chapter have been studied in a flow reactor used to measure the oxidation of CO by Williams and co-workers [82G01, 86L01]. As shown in Fig. 7.7 the effect of shock activation is substantial. Whereas the unshocked material displays such low activity that an effect could only be observed at the elevated temperature of 400 °C, the shock-modified powder shows substantially enhanced catalytic activity with the extent of the effect depending on the shock pressure. After a short-time transient is annealed out, the activity is persistent for about 8 h. Although the source of the surface defects that cause the activity is not identi-fied, the known annealing behavior of the point defects indicates that they are not responsible for the effect.

Similar studies have been carried out on shock-activated zinc oxide. Here, the effects are not so pronounced, but interesting effects are seen [86W04].

Dissolution

The high temperature α-β transformation of shock-activated silicon nitride powder has been investigated at temperatures of 1600 and 1700 °C [84B01].

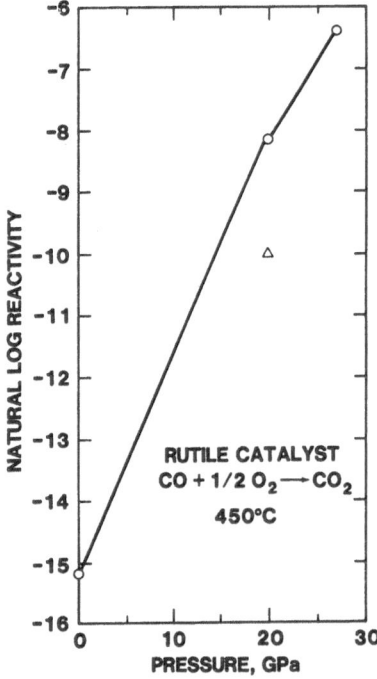

Fig. 7.7. Studies of the catalytic activity of shock-modified rutile in the oxidation of CO shows greatly enhanced catalytic activity, which is strongly influenced by the shock conditions [86G01].

The starting material and shock-activated powder were mixed with 5-wt% MgO and heated for various periods. At the end of each period the phase content of the samples was determined with x-ray diffraction. In this environment it is thought that the β phase is formed by a dissolution-precipitation process as shown in Fig. 7.8. As indicated in Fig. 7.9, the shock-activated silicon nitride displays substantially enhanced dissolution rates that are strongly dependent on shock pressure between 22 and 27 GPa.

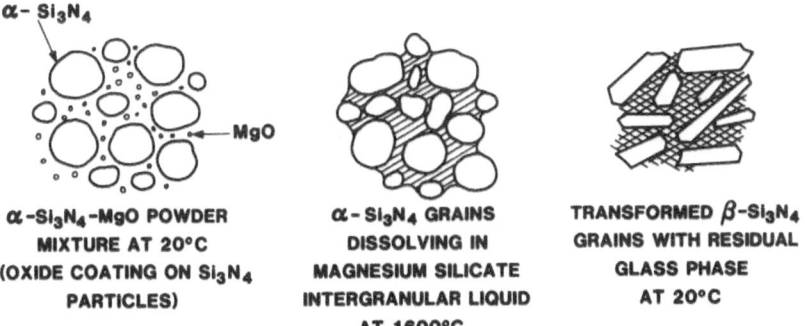

Fig. 7.8. High temperature conversion of α-silicon nitride with an MgO additive to the β-phase is thought to be a consequence of dissolution of the α phase in a magnesium silicate with subsequent recrystallization from the melt. Enhanced dissolution rate should then strongly influence α–β conversion [84B01].

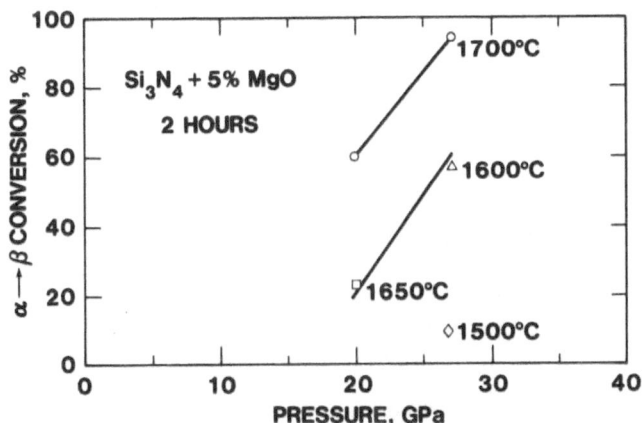

Fig. 7.9. Measurements of the degree of conversion of α–β silicon nitride at a fixed time and various temperatures are thought to show the strong influence of shock modification on the high temperature dissolution [84B01].

In a detailed study the dissolution kinetics of shock-modified rutile in hydrofluoric acid were carefully studied by Casey and co-workers [88C01]. Based on the defect studies of the previous sections in which quantitative measures of point and line defects were obtained, dissolution rates were measured on the as-shocked as well as on shocked and subsequently annealed powders. At each of the annealing temperatures of 200, 245, 330, 475, 675, 850, and 1000 °C, the defects were characterized. It was observed that the dissolution rates varied by only a factor of 2 in the most extreme case. Such a small effect was surprising given the very large dislocation densities in the samples. It was concluded that the dissolution rates were not controlled by the dislocations as had been previously proposed.

Solid State Chemical Reaction

Shock-modified zirconia powder was reacted with lead oxide in controlled differential thermal analysis (DTA) experiments and compared to the unmodified material by Hankey and co-workers [82H01]. This reaction yields

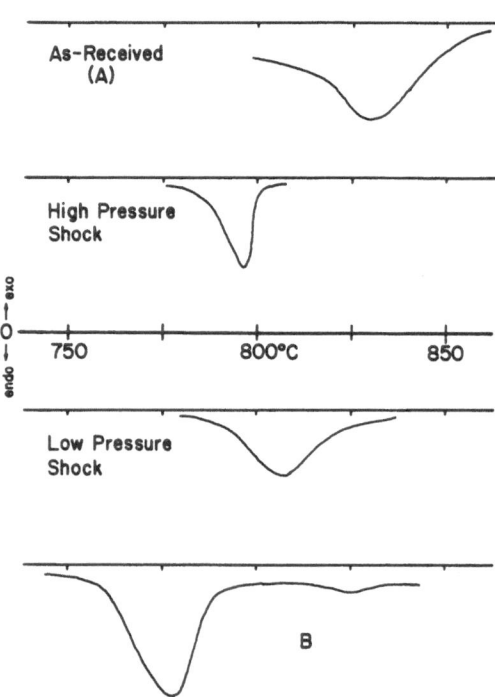

Fig. 7.10. The solid state reactivity of shock-modified zirconia with lead oxide as studied with differential thermal analysis (DTA) shows both a reduction in onset temperature and apparent increase in reaction rate. The shock-modified material has a behavior much like the much higher specific surface powder shown in "B" (after Hankey et al. [82H01]).

a lead-zirconate product with a significant endothermic energy change. The DTA traces for powder shock-modified at 22 and 27 GPa are shown along with the starting material and a higher specific surface material as shown in Fig. 7.10. The effect of the shock modification is to lower the reaction temperature as measured by the peak of the reaction curve. Although it is difficult to quantify, the shock-activated material shows a narrower response curve similar to the high reactivity material. Thus, the effect of the shock activation was to convert a zirconia of nominal reactivity to a material of significantly larger reactivity.

Proposed mechanisms for the observed reactivity enhancement include particle comminution, crystallite size reduction, retention of tetragonal phase zirconia, and recrystallization of heavily deformed monoclinic-phase material. The microstructure of the shock-modified zirconia was studied by Hellmann and co-workers [84H02] and it was observed that the 20 GPa shock-modified sample showed a high degree of deformation and crystallite size reduction. At higher pressure, the zirconia was characterized by recrystallization and localized particle size reduction accompanied by formation of tetragonal phase particles. Shock-formed tetragonal particles of about 30 nm in size were resistant to recrystallization in electron beam heating; however, heavily deformed shock-modified powders were observed to experience recrystallization with electron beam heating. Given the structural transformation measurements to be reported below, it seems likely that the enhanced solid state reactivity of zirconia in reaction with lead oxide is due to a combination of small tetragonal phase crystallites and reduced crystallite size.

Hot Pressing, Sintering

Early reports by Bergmann and co-workers and Anan'in and co-workers described enhanced sinterability in shock-modified refractory powders. Beauchamp [87B02] has reviewed the reported results and concluded that the observed effects are not due to defect structure, but more likely due to particle comminution and compaction of particles into high density conglomerates. To investigate the influence of the shock-formed defects, studies have been carried out on hot-pressing of aluminum nitride and aluminum oxide [87B03, 87B04].

Shock-modified aluminum nitride showed the largest effects, and microstructural studies were successful in identifying the source of significantly enhanced consolidation rates. As shown in Fig. 7.11 the density achieved after hot pressing at stresses of 47 and 26 MPa is observed to be significantly higher for shock-modified powders than for the same unmodified powders. The observed behavior was shown to be a consequence of dynamic recrystallization promoted by the high dislocation densities which resulted in enhanced dislocation motion. The model for the enhanced consolidation rates is supported by detailed microstructural studies of samples obtained in various stages of the densification process.

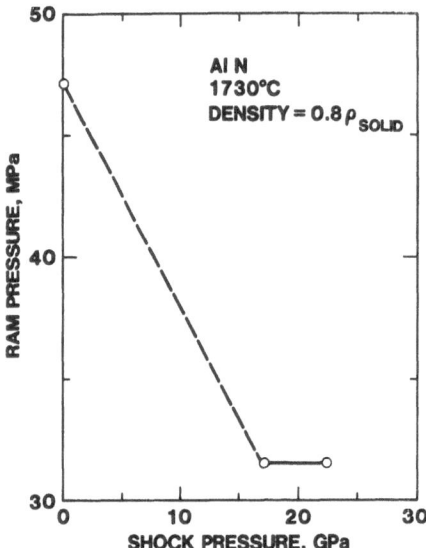

Fig. 7.11. The consolidation behavior in hot pressing of shock-modified AlN is found by Beauchamp and co-workers to be strongly influenced by shock modification [87B04].

Shock-modified aluminum oxide powders were studied in conventional sintering and hot pressing studies in an attempt to determine the cause of enhanced sinterability and to identify mechanisms for its reported occurrence. Beauchamp suggested that, at least in the early stages of sintering where densities are still low, the constraints on mechanical deformation are different from those at high density. If so, the shock-formed dislocations could give an enhanced deformation by dislocation slip in addition to the indirect contributions from enhanced diffusional processes. Hot pressing studies on aluminum oxide showed enhanced densification rates early in the process, but these effects were not persistent. The effect was thought to be the result of rapid sintering of agglomerates in the matrix of the more porous material. Because the regions of low porosity cannot sinter as rapidly as the agglomerates, they are eventually prevented from completing their densification because of the mechanical constraints of the high density zone.

Solid State Structural Transformations

The changes in structural transformation characteristics of shock-modified powders have shown pronounced effects, providing some of the most direct evidence for changes in materials characteristics resulting in enhanced solid state reactivity. Detailed DTA measurements on shock-modified zirconia for the thermally induced monoclinic to tetragonal transformation reveal significant effects [84H01]. Furthermore, thermal annealing of shock-modified, theta-phase aluminum oxide showed strong evidence for shock-induced formation of nucleation sites of the alpha phase [90B01].

Shock-modified zirconia was studied in DTA experiments to temperatures of 1500 °C [84H01]. As shown in Fig. 7.12, the temperature for transforma-

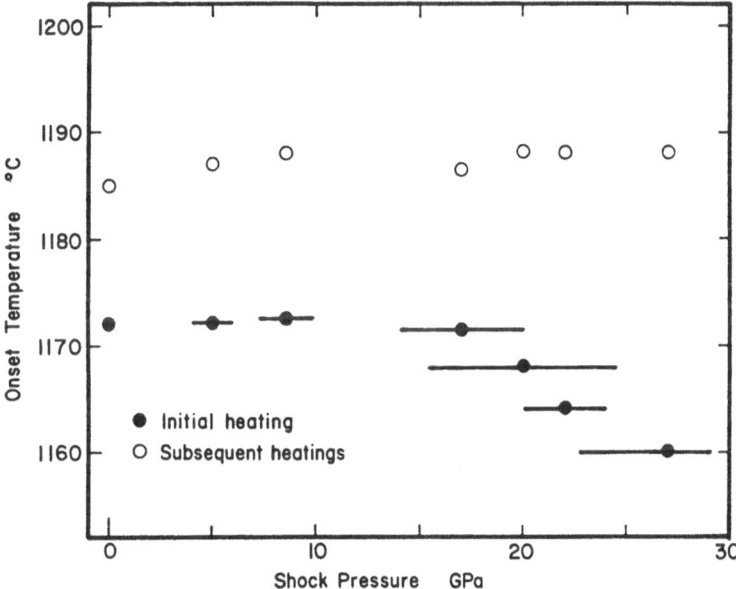

Fig. 7.12. The monoclinic to tetragonal conversion of shock-modified zirconia was studied with DTA by Hammetter and co-workers. The conversion temperature was found to be strongly changed and dependent on shock-modification conditions. The higher-pressure behavior was found to be strongly correlated with reduction in crystallite size [84H01].

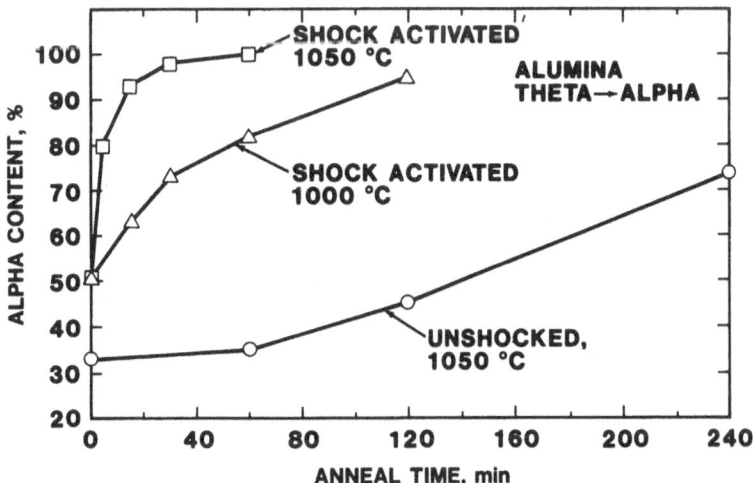

Fig. 7.13. The conversion of theta- to alpha-phase alumina was found to be strongly affected by shock modification in work of Beauchamp and co-workers [90B01]. Whereas the unshocked powder showed evidence for an incubation period of 60 min, the shock-modified materials show immediate conversion typical of the presence of shock-formed nuclei.

tion from the monoclinic to tetragonal phase for the unshocked powder is determined to be about 1185 °C. Shock-modified powders, which were treated at shock pressures from 5 to 27 GPa, were shown to exhibit reductions in transformation temperatures from 120 to 250 °C. For shock pressures up to 16 GPa the reductions were not shock-pressure dependent, with transformation temperatures reduced by about 120 °C. For shock-modified powders shock treated above 17 GPa, the transformation temperature was further reduced with increasing shock pressure by another 130 °C to the maximum pressure. The observed higher pressure effect was well correlated with crystallite size reduction.

Thermal treatment of shock-modified theta-phase alumina, which initially contained about 30% alpha phase, showed a dramatic change in the rate of transformation to the alpha phase [90B01]. As shown in Fig. 7.13, the shocked sample showed no evidence for an incubation period and displayed a rapid conversion to the alpha phase, in sharp contrast to the unshocked sample. Such behavior clearly indicates that the shock process resulted in formation of larger concentrations of alpha-phase nuclei.

Enhanced Solid State Reactivity

Shock-processed materials are so strongly modified in different aspects that it has proven difficult to identify mechanisms associated with enhanced solid state reactivity in terms of specific defects and structural features. Nevertheless, shock-modified powders have been observed to show widespread evidence for substantial shock activation. The particular influences depend strongly on the specific chemical process considered and on the detailed characteristics of the powder in its shock-modified state. In most cases, it is clear that significantly enhanced solid state reactivity is an expected consequence of shock treatment. The studies show that the morphology and solid state reactivity of reactants in potential shock-synthesis processes are qualitatively different from their starting condition at the time of reaction. A summary of enhanced solid state reactivity in shock-modified materials is given in Graham and Thadhani [93G01].

CHAPTER 8

Solid State Chemical Synthesis

In this chapter: synthesis of zinc ferrites, intermetallic compounds, and metal-oxides.

The National Research Council Report "Opportunities in Chemistry" published in 1985 (called the "Pimentel Report") described a survey of the state of science and technology in chemistry and made recommendations concerning future opportunities. One of the recommendations proposed the following:

"An initiative to explore chemical reactions under conditions far removed from normal conditions. Chemical behaviors under extreme pressures, extreme temperatures, ... provide critical tests of our basic understandings of chemical reactions and new routes toward discovery of new materials and new devices."

Certainly the ability to carry out chemical reactions at pressures of tens of GPa in times of a microsecond or less provides an opportunity to explore chemistry under extreme conditions as described by the report. Nevertheless, the conditions under which chemical reactions proceed under high pressure shock compression are complex and not well understood, and careful, systematic investigations are required to develop the ability to control and predict the reactions. The particular emphasis of work reported in this chapter is to study the solid state aspects of such processes.

The principal features of shock-induced solid state chemical processes were outlined in Chap. 6 along with the sample preservation technology. Based on the capability to subject powder compacts to controlled shock loading and preserve the samples for post-shock analysis, numerous characteristics of shock-modified inorganic powders were determined, as summarized in Chap. 7. Both physical and chemical characteristics have been determined. The work reported includes a sufficient number of different materials whose properties were investigated over a wide range of shock conditions to provide a broad base of observation for overall conclusions on modification and reactivity to be drawn. The materials property data provide a basis on which realistic descriptions can be given for reactants as they actually enter into a solid state chemical process. It is apparent from those data that the starting powder mixtures placed in a sample capsule for shock-induced reactions must be regarded as potential reactants that must undergo modification and activation before they can become involved in shock-induced chemical reaction.

There are numerous observations of shock-induced chemical reactions [83G02] in powder mixtures, but most of the investigations have subjected the materials to uncertain conditions that are unsuitable for exploration of the mechanisms and materials characteristics controlling the processes. As was the case in chapters describing shock modification and shock activation, the information presented in this chapter is limited to observations by the author and his co-workers. The degree of consistency achieved in the coordinated program investigating modification, activation, and chemical synthesis brings a substantial degree of confidence to the conclusions concerning mechanisms. In this chapter, chemical synthesis studies in mixed-oxide systems and intermetallics are described. Finally, observations on metal-oxide mixtures with large heats of reaction are described. Work on shock-induced chemical synthesis has been reviewed by Graham and co-workers [86G01, 86G02, 88G01, 89G01].

8.1 Zinc Ferrite Synthesis

The reaction of mixed powders of zinc oxide and hematite (Fe_2O_3) to form a zinc ferrite is a particularly interesting system for the study of solid state reactions. The selection of this reaction was based upon identification of the reaction as a typical solid state reaction which has been investigated in earlier, thermal synthesis experiments. Ferrites as a class have been extensively studied, and the relationships between magnetic properties at the macroscopic level are well connected to atomic-level physical properties. The early shock-synthesis work of Kimura [63K03] demonstrated that the reaction could be induced by a shock process. Both well characterized zinc oxide and hematite could be readily obtained, and the properties of both the starting hematite and the product, zinc ferrite, could be studied in detail with magnetization and Mössbauer techniques in addition to the more conventional x-ray diffraction probe. The reaction is accomplished with little or no heat of reaction so that thermochemical complications are not prominent in interpretations of the process. Without a heat of reaction, the products are not melted. This aspect is particularly important as melting of the product has the effect of removing evidence for material condition existing prior to the reaction. Furthermore, mechanical characteristics of the reactants are felt to be important, and these powders represent a soft, readily deformed powder (zinc oxide) with a hard, high strength powder (hematite). The densities of the reactants are also quite similar, so that influences of different localized particle velocities at fixed pressure are minimized. All of these features make this reaction an ideal vehicle for a study of shock-induced solid

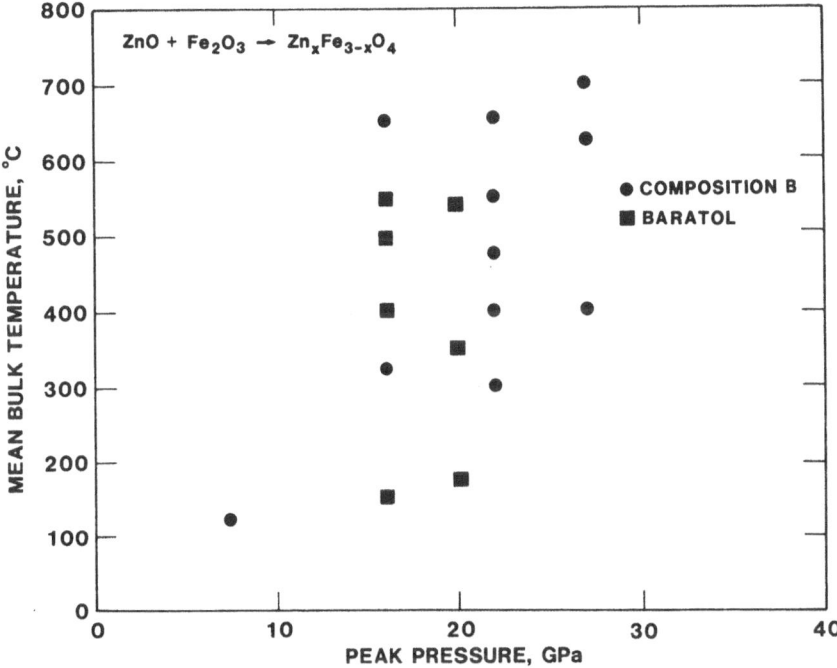

Fig. 8.1. Shock-induced solid state chemical synthesis of a zinc ferrite has been studied over a wide range of temperature and pressure. The figure shows the location of conditions for which the reaction has been studied.

state chemistry. The work has been reported in a number of individual papers [86M01, 86V01, 86W02, 86C01, 86G01].

Shock-synthesis experiments were carried out over a range of peak shock pressures and a range of mean-bulk temperatures. The shock conditions are summarized in Fig. 8.1, in which a marker is indicated at each pressure-temperature pair at which an experiment has been conducted with the Sandia shock-recovery system. In each case the driving explosive is indicated, as the initial incident pressure depends upon explosive. It should be observed that pressures were varied from 7.5 to 27 GPa with the use of different fixtures and different driving explosives. Mean-bulk temperatures were varied from 50 to 700 °C with the use of powder compact densities of from 35% to 65% of solid density. In furnace-synthesis experiments, reaction is incipient at about 550 °C. The melt temperatures of zinc oxide and hematite are >1800 and 1.565 °C, respectively. Under high pressure conditions, it is expected that the melt temperatures will substantially increase. Thus, the shock conditions are not expected to result in reactant melting phenomena, but overlap the furnace synthesis conditions.

Four different material probes were used to characterize the shock-treated and shock-synthesized products. Of these, magnetization provided the most sensitive measure of yield, while x-ray diffraction provided the most explicit structural data. Mössbauer spectroscopy provided direct critical atomic level data, whereas transmission electron microscopy provided key information on shock-modified, but unreacted reactant mixtures. The results of determinations of product yield and identification of product are summarized in Fig. 8.2. What is shown in the figure is the location of pressure, mean-bulk temperature locations at which synthesis experiments were carried out. Beside each point are the measures of product yield as determined from the three probes. The yields vary from 1% to 75 % depending on the shock conditions. From a structural point of view a surprising result is that the product composition is apparently not changed with various shock conditions. The same product is apparently obtained under all conditions; only the yield is changed.

Although good consistency on the characteristics of the products and how

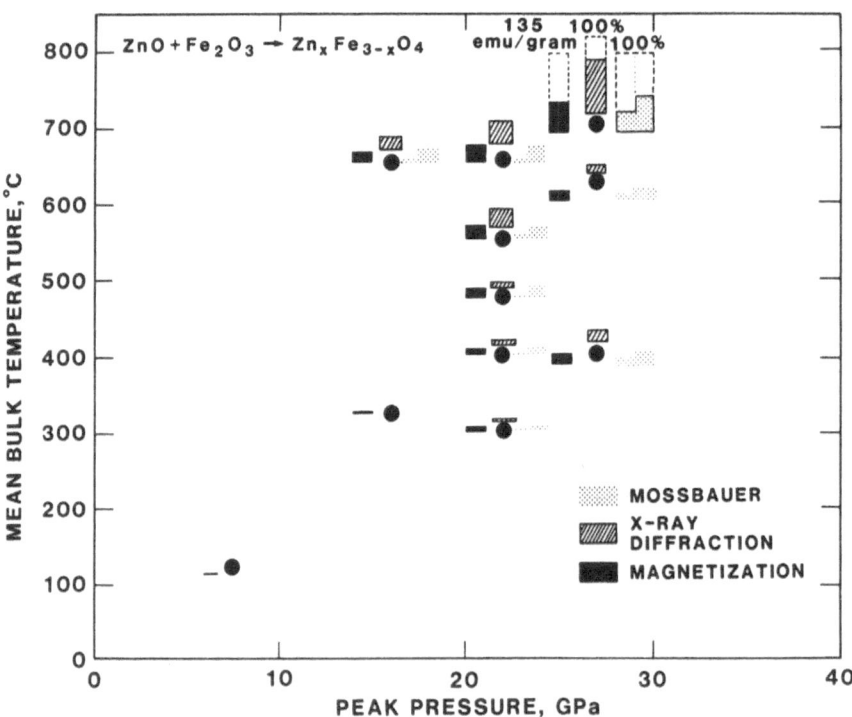

Fig. 8.2. The characteristics of shock-synthesized zinc ferrite are found to be sensitive to both shock pressure and temperature. The description of the product depends strongly on the materials probe used for analysis (after Graham et al [86G01]).

they change with shock conditions is obtained with any particular probe, it is clear that each probe provides a somewhat different description of the product. Magnetization measurements show the yield of product as a continuous function of shock pressure and temperature; even at the most modest loading, chemical reaction is observed in small yields. X-ray diffraction gives unequivocal data identifying the spinel structure of the product, but cannot differentiate between products of the same structure in different magnetic states. The Mössbauer data are particularly revealing as they clearly show two products with different local magnetic characteristics.

The spectra clearly show iron in three distinct resonances. One of these resonances is due to iron in the shocked but unreacted hematite, and its characteristics match those of hematite as described in Sec. 7.1. Thus, the Mössbauer data are explicit in demonstrating that changes in magnetic properties are due to the reaction products, not some defect-induced modification of hematite. The second iron resonance is identified as a zinc-deficient ferrite. The third iron resonance is associated with paramagnetic material. This paramagnetic phase is apparently of the same structure as the magnetic phase and most likely represents material in a superparamagnetic phase resulting from small, mesoscopic scale crystallite size.

Study of a shocked, but modestly reacted, sample with transmission electron microscopy was revealing as to the morphology and degree of mixing of the reactants prior to reaction. Both reactants were shown to be of low defect density, and no reaction products could be identified. The principal feature of the samples was that only hematite could be easily identified. The softer zinc oxide had flowed within and between the hematite grains. X-ray energy dispersion spectroscopy (EDS) across grain boundaries showed heavy concentrations of zinc oxide; up to 5% zinc oxide was found within the hematite grain. The substantial shock-induced modifications provide direct evidence for mechanically induced "mixing." No evidence was seen for localized melting.

In spite of careful analysis of the products with the various sophisticated probes, differences in the composition are reported. All measurements indicate a zinc-deficient zinc ferrite, but the indicated zinc concentration varies from 0.2 to 0.9. The EDS measurements are based on direct zinc concentration measurements. Determinations based on magnetization and Mössbauer spectra are obtained on zinc ferrite synthesized in more conventional processes.

One of the most interesting results of the zinc ferrite synthesis is the observation that the yield of the product is dependent on the early pressure history. This behavior is shown in Fig. 8.3, which plots the yield versus temperature for baratol explosive loading and for Composition B explosive loading. The difference between these loadings is that the initial pressure pulse amplitude is significantly greater with Composition B. Apparently, the early pressure history has an important conditioning effect for subsequent reactions.

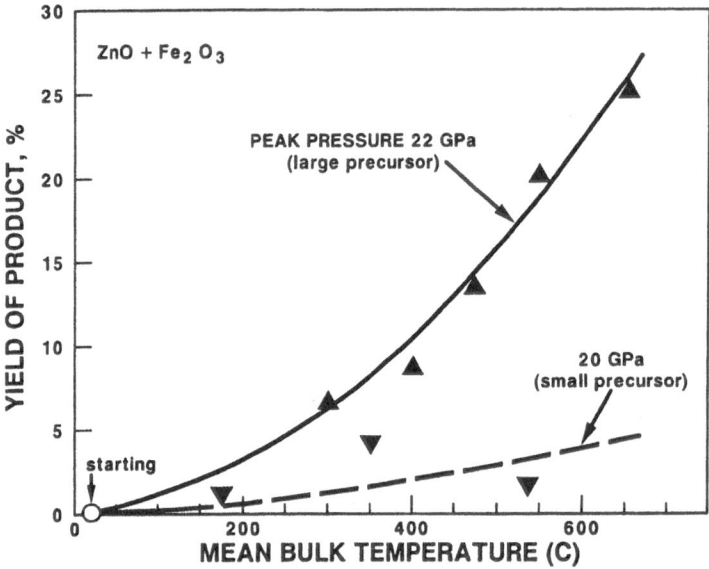

Fig. 8.3. The yield of shock-synthesized zinc ferrite is found to be strongly dependent on the early loading history. This characteristic is thought to be an indication of shock modification on subsequent chemical reaction.

8.2 Intermetallic Compound Synthesis

Nickel Aluminides

Intermetallics also represent an ideal system for study of shock-induced solid state chemical synthesis processes. The materials are technologically important such that a large body of literature on their properties is available. Aluminides are a well known class of intermetallics, and nickel aluminides are of particular interest. Reactants of nickel and aluminum give a mixture with powders of significantly different shock impedances, which should lead to large differential particle velocities at constant pressure. Such localized motion should act to mix the reactants. The mixture also involves a low shock viscosity, deformable material, aluminum, with a harder, high shock viscosity material, nickel, which will not flow as well as the aluminum.

The reaction is significantly exothermic with a heat of reaction of about 40 kcal mol^{-1}. This energy will produce a sufficiently high temperature to melt the product and will allow the influence of thermochemical factors to be investigated. The temperature required to initiate the Ni–Al reaction at atmospheric pressure is about 660 °C. This reaction temperature threshold will be encountered in the shock processing, but it should be recognized that the conventional synthesis process is preceded by melting of the aluminum. At the pressure of the shock compression, the melt temperature of the aluminum will be approximately doubled to a value above the mean-bulk temperature encountered. Upon release of pressure it may be possible to melt the aluminum.

Two independent, but coordinated efforts on nickel–aluminum synthesis have been carried out. The first was by Horie and co-workers [85H01, 86H01], and the second by Thadhani and co-workers [89T01, 91D01]. In both cases the Sandia Bear fixtures were used for sample preservation. The investigations included the study of threshold conditions to initiate reactions, the effects of morphology, starting materials ratios, and density. Recent work [91S03] on the effect of powder compact density on initiation of reaction reveals the critical nature of the configuration of pore volume on the process.

The synthesis effort was initiated by the Horie group on mechanically blended powder mixtures of 3 parts nickel with 1 part aluminum in molar proportions and a similar sample composed of a composite particle of nickel plated on aluminum in similar proportions. The powders were a 44–74 μm nickel powder and a 5–15-μm micron aluminum powder, a coarse–fine mixture. The powder mixtures were shock loaded to peak pressures of 7.5 and 22 GPa with starting powder densities of 60% of solid density.

At the lower pressure, the sample is observed to be unreacted but significantly modified. It is found that the harder nickel particles largely retained their approximately spherical shape. The aluminum, on the other hand, is observed to flow between the nickel particles, and original particle morphologies are not preserved. Although the nickel was not deformed in external shape, etching of the particles showed evidence for recrystallization characteristic of recovery from dense concentrations of dislocations. There is evidence for nickel mixing into the deformed aluminum over distances of the order of particle dimensions. No evidence is found for localized melting.

At the higher pressure a distinctive reaction product is observed. As shown in Fig. 8.4, a triangular block of Ni_3Al is found in the outer edge of the sample. It should be remembered that numerical simulation predicted a temperature about 300 °C higher than in the bulk of the capsule. Within the reacted region, the product Ni_3Al is clearly identified. The product shows evidence of cooling from a melt and contains numerous voids. If the reaction occurred rapidly, temperatures of several thousands of degrees would have been produced. In the regions immediately surrounding the fully reacted sample, a gradient of reaction products is observed, including NiAl and Ni_2Al_3.

In the bulk of the sample, removed from the high temperature region, the powders show evidence for various degrees of sparse reactions; i.e., isolated local reactions. The regions are generally too small to be identified with x-ray diffraction but they can be identified from colors and appearances of etched surfaces. As at lower pressure, there is strong evidence for high concentrations of nickel in the deformed aluminum which can be loosely described as a solid solution of NiAl and Ni_3Al regions a few microns in size. Because of their small size, the thermal cooling from surrounding material occurs in a few nanoseconds. Under these conditions the reaction would only proceed to completion if the reaction rate led to complete reaction in a few nanoseconds.

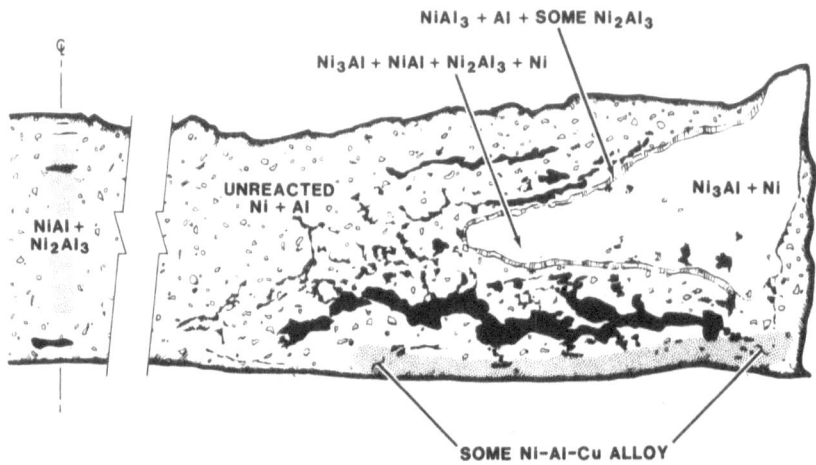

Fig. 8.4. A cross section of a shock-treated, Ni–Al powder mixture compact shows a characteristic reaction in the hotter, outer portion of the compact. There is also a localized reaction in the axially focused region [85H01].

Products of nickel aluminides are also observed along the axis of the sample over a diameter of about 1 mm where the numerical simulations show radial focusing of the pressure to values approaching 50 GPa for times of about 200 ns.

In perhaps the most important technological observation, it was found that the reacted Ni_3Al products have hardness values equivalent to cold-rolled or rapidly solidified Ni_3Al containing boron additives. The hardness is notable in that the material contains a low density of dislocations. It is observed that crystallites are less than 10 nm in dimension.

The strong influence of morphology and mixing is well illustrated with the composite particle investigation. These particles were composed of a nickel shell coated on spherical aluminum particles by hydrogen reduction in aqueous metal salt solution. The overall ratio of material in a particle was about 80 wt% Ni and 20 wt% aluminum. With these particles, the ratio of reactants was approximately the same as in the mixed powders, but the morphology of the reactants is radically different.

Under the higher pressure and temperature in which the mixed powders were found to be fully reacted, the composite powders were found to show no evidence for reaction over significant volumes of materials. Rather, it is observed that the particles are substantially deformed, principally flattened in the initial shock direction, with the nickel completely surrounding the deformed aluminum cores. The optical micrographs show that the morphology prevented significant mixing between the reactants. Localized reactions are observed at the interface between the nickel and aluminum but they are very limited in extent. No evidence is found for localized melting.

Thermal Analysis

Perhaps the most definitive result to come from the early nickel–aluminide synthesis work was the thermal analysis investigation of Hammetter [88H01, 88W01], which showed explicit data on substantial changes in the shocked-but-unreacted mixtures. Differential thermal analysis was carried out on the starting powder compacts of both the mechanically mixed and composite powders. Shocked and unreacted powders were compared to provide direct evidence for substantial changes introduced by the shock process.

The unshocked powders showed the initiation of reaction at a temperature

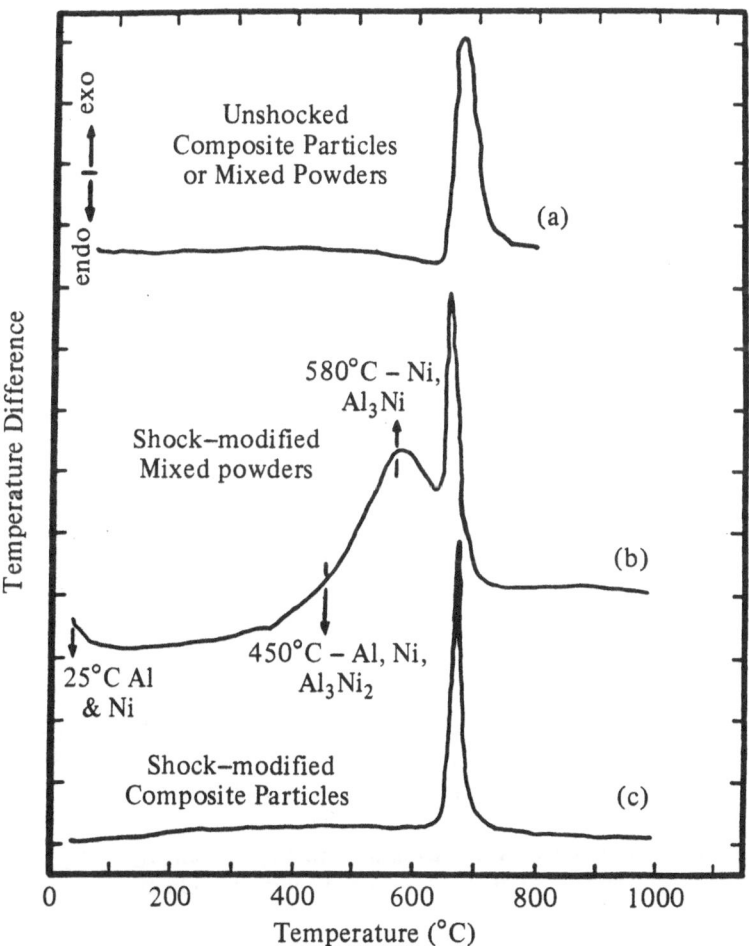

Fig. 8.5. The exothermic energy release of shocked but unreacted Ni–Al mixtures shows a profound change. A "preinitiation" phenomenon in which reaction temperature is reduced by over 200 °C is caused by the shock process. A compact composed of composite particles that inhibit mixing shows no such effect [88H01].

SHOCK–MODIFICATION CAUSES EXTENSIVE
MIXING BETWEEN THE A*l* AND Ni POWDERS

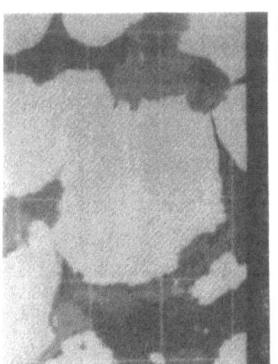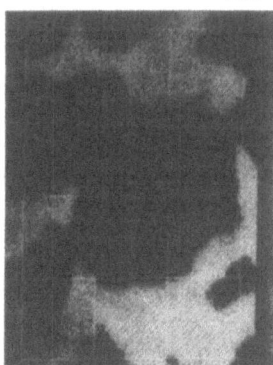

Fig. 8.6. Elemental distribution maps of a shock modified but unreacted powder mixture in the work of Hammetter et al. [88H01] show that there is considerable mixing of nickel into the aluminum.

of 650 °C with the composite particle showing the best signature due to the intimate interface achieved between the reactants. The DTA traces of the starting powders and the shock-modified powders are compared in Fig. 8.5 to illustrate the preinitiation phenomena prominent in the shock-treated powders. These shock-treated powders show a strong exothermic event initiating 250 °C below that for the unshocked powders. Samples taken at various stages of reaction identify the products as different from the final product of Ni_3Al.

The shock-modified composite nickel-aluminide particles showed behavior in the DTA experiment qualitatively different from that of the mixed-powder system. The composite particles showed essentially the same behavior as the starting mixture. As shown in Fig. 8.5 no preinitiation event was observed, and temperatures for endothermic and exothermic events corresponded with the unshocked powder. The observations of a preinitiation event in the shock-modified mixed powders, the lack of such an event in the composite powders, and EDX (electron dispersive x-ray analysis) observations of substantial mixing of shock-modified powders as shown in Fig. 8.6 clearly show the first-order influence of mixing in shock-induced solid state chemistry.

Powder Morphology Effects

The initial studies on nickel-aluminide synthesis defined a number of important issues in shock-induced solid state synthesis. This work was extended to the influence of powder particle morphology in recent work of Thadhani and

Table 8.1. Summary of aluminum–nickel mixture experiments.

Experiment No.	Powder morphology	Peak pressure (GPa)	Density (% solid)
Initial investigation 3Ni + Al—Horie et al. [86H01]			
93G846	coarse–fine	7.5	62
71G846	coarse–fine	16	62
72G846	coarse–fine	22	62
08G856	coarse–fine	27	62
Morphological effects 3Ni + Al—Thadhani et al. [91D01]			
38G876	fine–fine	22	61
39G876	coarse–coarse	22	61
40G876	flaky–coarse	22	62
41G876	coarse–fine	22	61
01H896	flaky–coarse	16	60
02H896	flaky–coarse	7.5	58
93H896	fine–fine	16	59
Reactant ratio effects Ni:Al—three morphologies—Thadhani et al. [91D01]			
21H896	3:1 coarse	22	61
22H896	3:1 fine–fine	22	61
23H896	3:1 flaky–fine	22	60
24H896	1:1 fine–fine	22	65
31H896	1:1 flaky–fine	22	67
36H896	1:1 flaky–fine	22	67
29H896	1:1 coarse–fine	22	67
25H896	1:3 flaky–fine	22	66
26H896	1:3 fine	22	65
30H896	1:3 coarse	22	67
37H896	1:3 coarse	22	67
Composite particles—Horie et al. [86H02]			
01G856	80Ni + Al	22	62
07G856	80Ni + Al	16	62

his co-workers [89T01]. As shown in Table 8.1, shock-synthesis experiments were carried out on four different aluminum–nickel powder mixtures, all shocked in starting material mixtures of the Ni_3Al stoichiometry. The mixtures included fine–fine, fine–coarse, medium–medium, and flaky–medium compositions. These powder mixtures were shock treated in the same conditions as for the initial studies.

Substantial differences were observed in the reactions. The fine–fine and flaky–medium mixtures were observed to be fully reacted under conditions in which reaction was observed only in the elevated temperature region in the outer edge of the samples. Thus, the threshold shock pressures or temperatures of the reactions were observed to be substantially lowered with these morphologies.

The unreacted samples and starting powder mixtures were studied in DTA experiments to investigate the effects of morphology on the shock modifica-

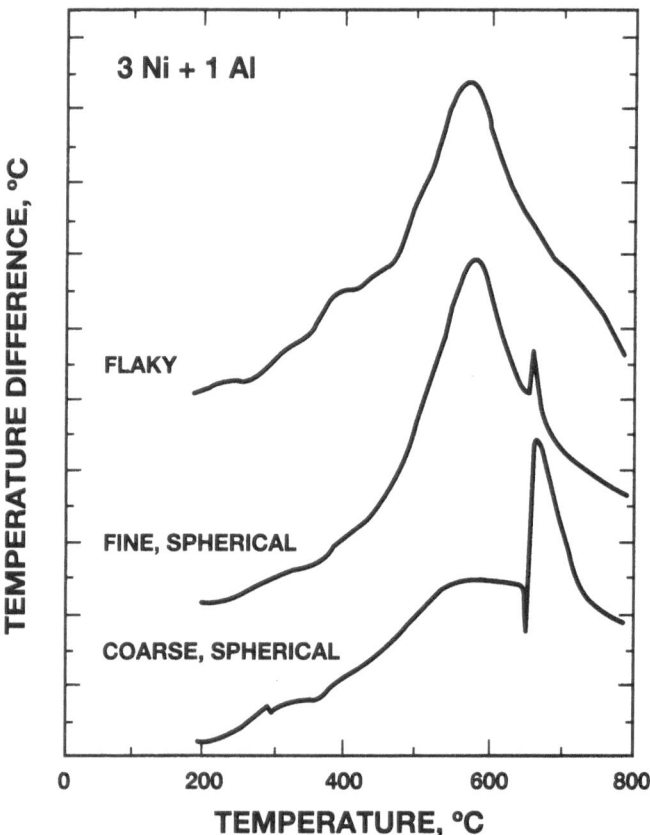

Fig. 8.7. The influence of powder morphology (configuration) on shock modification controlling initiation of reaction is shown by the thermal response of mixed Ni–Al powders of different morphologies. The preinitiation event shown in Fig. 8.5 is observed to be strongly influenced by morphology at fixed shock condition. The coarse–medium mixture shows the largest preinitiation event [91D01]. The data show much larger preinitiation events for the flaky and fine morphologies.

tion [91D01]. The DTA traces for the shock modified but unreacted samples show the preinitiation event, confirming the earlier work and providing new evidence describing the morphological effects. A summary of the DTA data is shown in Fig. 8.7. The preinitiation event is substantially larger in the more reactive mixtures. Unlike other morphologies, the flaky–coarse samples shocked at peak pressures of 7.5 GPa were found to be unreacted.

Reactant Ratio Effects [91D01]

Even though powders may be mixed in certain starting ratios, there is no fundamental reason why the reaction product should correspond to the starting ratio. Before reaction can occur, a mixing process must be accom-

Table 8.2. Molar and volumetric ratios of nickel–aluminum powders.

Molar	Weight	Volume
3Ni + 1Al	87:13	66:34
1Ni + 1Al	69:31	40:60
1Ni + 3Al	42:58	18:82

plished; this mixing process can be influenced by the ratio of reactants. Furthermore, chemical reaction processes may be more favorable at ratios that differ from the stoichiometric product. The effects of various starting ratios in the nickel-aluminide system are investigated under the same shock-loading conditions as in the earlier work for reactant ratios of 2:1, 1:1 and 1:2 compared to the earlier work on 3:1 mixtures. (The notation is Ni:Al.) As the influence of mixing is determined by the volumetric ratios of the potential reactants (starting materials), that ratio is shown in Table 8.2 for the present mixes.

The samples show strong effects on both threshold conditions to initiate reaction and reaction product. Principal differences are seen in the coarse powder mixtures in which significant reactions are observed in 3:1 mixtures. Mixtures of 1:3 show no reaction due to a lack of mixing. Figure 8.8 shows the DTA traces for the coarse powders in the various starting material ratios. Clearly, the preinitiation behaviors are strongly influenced by the starting material ratios. This observation, and the others shown above for the nickel aluminides, provides explicit support for the first-order influence of mechanical mixing in the control of shock-induced solid state chemical processes.

Other Intermetallics [86H01, 89T01]

The response of titanium–aluminum powder mixtures in a 3:1 molar ratio was investigated under the same shock-loading conditions as in the nickel aluminides. Such mixtures are especially interesting in that the shock impedances of the materials are approximately equal and both are relatively hard and difficult to deform. In addition to any chemical differences, such materials should prove to be difficult to mix with the shock conditions.

It was observed, under conditions when the nickel-aluminide mixtures of the same ratio were fully reacted, that the titanium aluminides were essentially unreacted; reactions were only localized. Because the products were of such small size, it was difficult to identify them, but they were thought to be $TiAl_3$ or ordered superstructures Ti_9Al_{23} or Ti_8Al_{24}. No further studies have been carried out on these samples.

Very high pressure and temperature experiments with the Sawaoka fixture on Nb–Si powder mixtures show that the silicon melted but the higher melt temperature niobium did not. Under these conditions, only chemical reaction

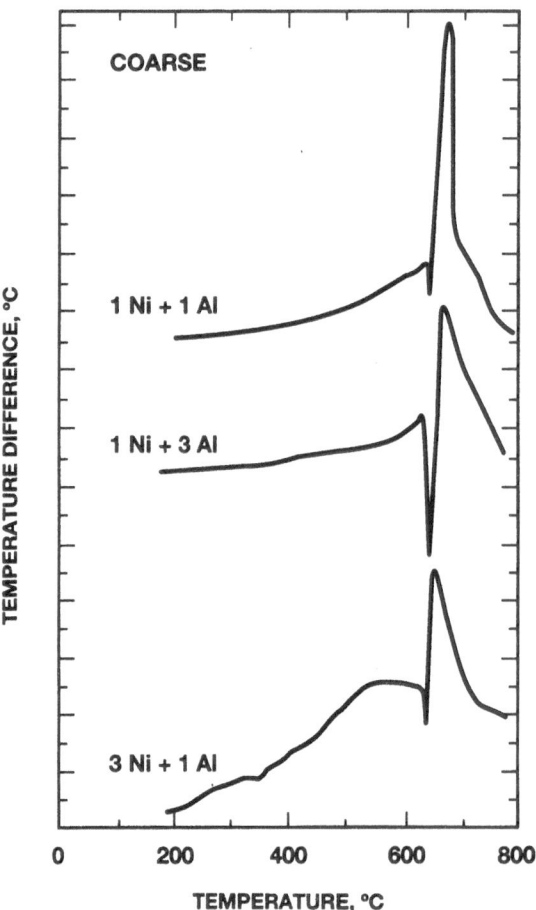

Fig. 8.8. The influence of starting material ratio on shock conditioning has been investigated for coarse powders in ratios of nickel to aluminum of 3:1, 2:1, and 1:1. The data show a strong influence of the ratio of the potential reactants consistent with the concept of mechanical mixing. [91D01].

localized on the Nb surface was observed; there was little mixing between the reactants. This observation further emphasizes the importance of shock-compression mixing due to plastic deformation. Solid particles in a liquid matrix will be subjected to limited plastic deformation due to the hydrostatic pressure field of the liquid.

8.3 Metal-Oxide Systems

The reactant systems considered above produce products that result in significant but limited heats of reaction. Given the significant temperatures that can be produced by the thermochemical contribution [92B02] it is of interest

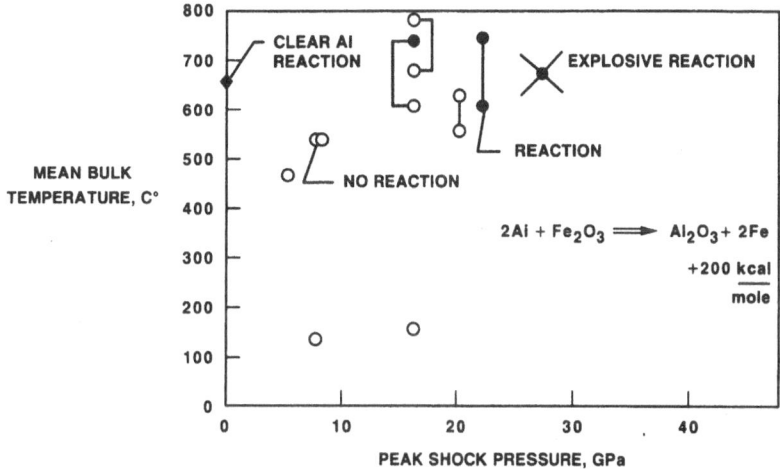

Fig. 8.9. The strongly exothermic reaction of hematite and aluminum mixtures shows effects strongly dependent on shock conditions that vary from no reaction to a strong, vigorous reaction. The observed behavior indicates that the heat of reaction does not play a dominant role in initiation of reaction.

to investigate a system with a very large heat of reaction. The reaction of $2Al + Fe_2O_3 \rightarrow Al_2O_3 + Fe$ has a heat of reaction of about 200 kcal mol^{-1}. Such a thermochemical contribution will cause temperatures of thousands of degrees. As in other investigations, the Sandia Momma Bear fixtures were used to preserve shocked samples over a wide range of peak shock pressure and mean-bulk temperature as controlled by the density of the powder compact.

The results of the study are shown in Fig. 8.9, in which points are shown at pressures and temperatures for which experiments were conducted. The samples show evidence for conditions in which no reaction is observed to conditions in which the reaction was so violent that no sample was recovered.

Of particular interest are the two samples at a pressure of 17 GPa. Each of the two experiments is shown by two points. The upper temperature points are the temperatures of the outer, hotter edge of the sample, and the lower points are those of the bulk of the sample. One of the samples shows no evidence for reaction: even the nominal conditions are the same. This sample was shock loaded with baratol explosive, which produces an early loading history lower in pressure than a sample with Composition B explosive loading. Apparently, the reaction is sensitive to the early shock conditioning. The upper point shows that the sample reacted in the outer region and not in the bulk of the sample. Actually, the copper plug in contact with the rear surface shows that a high temperature was achieved in the outer region and not in the bulk region. Examination of the sample as recovered after the experiment shows that it is fully reacted. Apparently, this sample provides an example of a case in which the reaction occurred during the shock event in the outer region, and this reaction caused a thermally propagating reaction into the bulk of the sample.

Hammetter [91H01] has carried out DTA analysis of a shock-modified but unreacted $2Al + Fe_2O_3$ mixture and found a preinitiation event, indicating the presence of shock-produced Hercynite (spinel phase $FeAl_2O_4$). These preinitiation events observed in thermal analysis provide direct evidence for the influence of mechanical, shock-induced mixing in controlling solid state chemistry.

8.4 Shock-Induced Solid State Chemical Synthesis

This chapter presents detailed and thorough studies of chemical synthesis in three quite different chemical systems: zinc ferrite, intermetallic, and metal oxide. In addition to different reaction types (oxide–oxide, metal–metal, and metal oxide), the systems have quite different heats of reaction. The oxide–oxide system has no heat of reaction, while the intermetallic has a significant, but modest, heat of reaction. The metal oxide system has a very large heat of reaction. The various observations appear to be consistent with the proposed conceptual models involving configuration, activation, mixing, and heating required to describe the mechanisms of shock-induced solid state chemistry.

The observed behaviors show that shock-induced mechanical mixing is probably the key process controlling rate of reaction. Clear evidence was found for shock mixing in shocked-but-unreacted powder mixtures. These samples show potential reactants mixed to various degrees. It appears likely that in every case there is a statistical distribution of reactants produced and controlled by shock-induced mechanical mixing. No doubt, reaction initiates at localized locations in which combinations of reactants are raised to an appropriate temperature. As the key factor in solid state chemistry, mixing requires intense study if quantitative modeling is to be accomplished. The observations indicate that heat of reaction has little, if any, influence on initiation of reaction. To the extent that viscoplastic deformation is the controlling factor leading to mixing, melting of one or both reactants can actually impede chemical reaction on a microsecond time scale.

Given the importance of mixing in controlling the reactions, it is apparent that, in most cases, the potential reactants are not all able to participate in the reaction. These inert species would be expected to play a critical role in thermal quenching of a local reaction. The VIR Model of Horie [90K01] provides a basis for describing such phenomena. First-order thermochemical models such as those of Boslough [92B02] and Yu and Meyers [91Y01] offer insight into the effect of solid state reactions on pressure and temperature.

PART V

Shock Compression of Solids as a Mechanical, Physical, Chemical Process

CHAPTER 9

The Shock-Compression Processes

In this chapter: solid-state physics and chemistry; future directions in shock-compression science.

9.1 Solid State Physics and Solid State Chemistry

Shock-compressed solids and shock-compression processes have been described in this book from a perspective of solid state physics and solid state chemistry. This viewpoint has been developed independently from the traditional emphasis on mechanical deformation as determined from measurements of shock and particle velocities, or from time-resolved wave profiles. The physical and chemical studies show that the mechanical descriptions provide an overly restrictive basis for identifying and quantifying shock processes in solids. These equations of state or strength investigations are certainly necessary to the description of shock-compressed matter, and are of great value, but they are not sufficient to develop a fundamental understanding of the processes.

The descriptions of shock-compressed solids resulting from the physical and chemical investigations lead to a picture significantly different from the mechanical investigations. In this book, the descriptions are developed from the most simple—the elastic compression range—to the most complex—solid state chemistry. Approaching the problem in this manner permits careful examination of fundamental assumptions and challenges to the picture developed from the mechanical studies. The overall picture that emerges is one in which shock-compressed matter and shock-compression processes are a great deal more complex than the mechanical measurements indicate. A picture of a homogeneous compression of solid lattices with some modification for the stress tensor to account for the solid state aspects is simply insufficient. It is excessively naive. From the physical and chemical property viewpoint, the influence of large concentrations of defects from atomic to mesoscopic to macroscopic dominate the description. Such considerations require anisotropic, heterogeneous conditions and require explicit descrip-

197

tions which are not in thermodynamic equilibrium. It should be recognized that anisotropic and heterogeneous conditions can be far more effective in influencing processes than isotropic and homogeneous conditions.

This book has certainly not included all observations of solid state physical and chemical processes that have been carried out over the past 40-yr history of shock compression science, but has necessarily emphasized those studies in which the author has made significant contributions. Other independent studies reinforce these observations.

It has been observed in moving from the simplest cases to the more complex cases that, even in the simplest situation of elastic shock deformation, the observed effects can be considerably more complex than would be anticipated from extrapolation from low pressure, small signal studies carried out in more conventional solid state physics. There is typically a degree of connection to the conventional work, but complexities, usually associated with defects, enter the picture from unanticipated directions.

The concept of a well defined elastic range to large strain is not realistic. The concept of well defined stress at which mechanical yielding occurs leading to well defined elastic–inelastic conditions is not realistic. Actually, such conclusions could well be anticipated from strength studies at atmospheric pressure, but there has been little explicit reason to consider the nonideal effects from the mechanical-response shock studies.

Given the complexities of the elastic range it is not surprising that the investigations in the elastic–inelastic range create conditions that are, in many cases, not subject to definitive analysis. Certainly, the viscoplastic aspects of shock deformation have been considered, but the descriptions are not sufficient to define physical and chemical processes. Nevertheless, if proper account is taken of deviations from simple models, the first-order behaviors are sometimes interpretable, especially in situations involving magnetism, which are insensitive to defects. In contrast, a problem involving semiconductors in the elastic–inelastic compression range is essentially uninterpretable due to the dominance of defects in influencing their physical behavior.

Shock phenomena, such as shock-induced polarization, have no known counterpart in other environments. In that regard, the distinctive behaviors present the greatest opportunity to determine details of shock-compression processes. Unexplored phenomena, such as electrochemistry [88G02], offer considerable potential for developing improved descriptions of shock-compressed matter.

Shock-induced solid state chemistry represents the most complex fundamental problem ever encountered in shock-compression science. All the mechanical and physical complications of other work are present, yet the additional chemical complications are added. Indeed, all mechanical, physical, and chemical aspects of the problem are intimately intertwined. Chemical investigations promise to provide a description of shock compression that differs considerably from that to which we have become accustomed. Nevertheless, a full description of the process requires contributions from a number of established scientific fields, including materials science, solid state physics, and solid state chemistry.

9.2 Future Directions in Shock-Compression Science

When the author came to the newly developing research group at Sandia Laboratories in 1958 with a background in solid mechanics, the immediate troubling question that emerged from study of the literature was how to understand the profound change of solids that resulted in their fluidlike deformation under high pressure shock compression. This book summarizes much of the intervening study undertaken to answer that question. Can the question be answered on the basis of the existing research?

There is a view developing concerning the accomplishments of shock-compression science that the initial questions posed by the pioneers in the field have been answered to a significant degree. Indeed, the progress in technology and description of the process is impressive by any standard. Impressive instrumentation has been developed. Continuum models of materials behavior have been elaborated. Techniques for numerical simulation have been developed in depth.

Nevertheless, given that the early questions have been answered to a considerable degree, have we been successful in answering the fundamental questions involved in shock compression? Have we been able to answer Bridgman's [56B01] concern that "...the precise mechanisms by which reaching the plastic flow point may induce the discontinuity (the 13 GPa transition) seem not to have been worked out?" We have certainly been successful in determining that we have a very limited mechanical, physical, and chemical understanding of shock-compressed matter. Nevertheless, it is difficult to see how further investigations along established directions will lead to an improved fundamental understanding.

After 40 yr of intensive research is it now time to assess the directions followed and what we have accomplished, time to recognize that we need to pose new questions and need fundamentally new probes to understand shock-compressed matter, and to critically examine the fundamental assumptions, implicit and explicit, that we have employed? Is it now time for a paradigm shift to a more realistic, if more complex, description of shock-compressed matter? Can we continue to view shock-compressed matter as an analog to a static high pressure that happens to be achieved in a short time and has a large thermal component attached?

Surely, it is now time to reformulate the questions considered to be fundamental to shock-compression science. The questions must consider shock-compressed matter as it exists as a highly defective solid, heterogeneous in character, with significant anisotropic components and heterogeneous processes that are not in thermodynamic equilibrium.

Major efforts have developed descriptions of shock processes with molecular dynamics [88B01], but these descriptions cannot be verified with experimental measurements. Such molecular level approaches must be integrated into descriptions at the mesoscopic level, at the microscopic level, and into continuum descriptions. Of course, the descriptions must realistically describe the material in its defect state. Perhaps new methods of describing plastic deformation as excited states analogous to solid state descriptions of electronic phenomena can pose quite different questions concerning disorder in lattice behavior and rotational components of material deformation at the continuum level [87E01, 87O01]. Incorporation of modern materials science technology into continuum descriptions remains an unsolved challenge.

It is appropriate today for those interested in shock-compression science to recognize the situation posed by the eminent Swedish solid state chemist, J. Arvid Hedvall, who called attention to the situation "of the hypnotic power of mental habitudes" [66H01].

References

20P01   C.A. Parsons, Philos. Trans. R. Soc. London, Sec. A **220**, 67 (1920).

48P01   D.C. Pack, W.M. Evans, and H.J. James, Proc. Phys. Soc. London, **60**, Pt. 1, 1–8 (1948).

51B01   R.M. Bozorth, *Ferromagnetism* (Van Nostrand, New York, 1951).

53P01   M.B. Prince, Phys. Rev. **92**, 681–687 (1953).

54M01   F.J. Morin and J.P. Maita, Phys. Rev. **94**, 1525–1531 (1954).

55D01   G.E. Duvall and B.J. Zwolinski, J. Acoust. Soc. **27**, 1054–1058 (1955).

55G01   R.W. Goranson, D. Bancroft, B.L. Burton, T. Blechar, E.E. Houston, E.F. Gittings, and S.A. Landeen, J. Appl. Phys. **26**, 1472–1479 (1955).

55H01   C. Herring, Bell Syst. Tech. J. **34**, 237–290 (1955).

55M01   H.D. Mallory, J. Appl. Phys. **26**, 555–559 (1955).

55M02   S. Minshall, J. Appl. Phys. **26**, 463–469 (1955).

55M03   S. Minshall, Phys. Rev. **98**, 271 (1955).

56A01   B.J. Alder and R.H. Christian, Disc. Faraday Soc. **22**, 44 (1956).

56B01   P.W. Bridgman, J. Appl. Phys. **27**, 659 (196).

56R01   I.N. Riabinin, Sov. Phys. Dokl. **1**, 424–426 (1956).

57A01   G.W. Anderson and F.W. Neilson, Bull. Amer. Phys. Soc. **2**, 302 (1957).

57N01   F.W. Neilson, Bull. Am. Phys. Soc. **2**, 302 (1957).

58A01   L.V. Al'tshuler, K.K. Krupnikov, B.N. Ledenev, V.I. Zhuchikhin, and M.I. Brazhnik, Sov. Phys. JETP **34**, 606–614 (1958).

58G01   R. Grover, R.H. Christian, and B.J. Alder, Bull. Am. Phys. Soc. **3**, 230 (1956).

58R01   M.H. Rice, R.G. McQueen, and J.M. Walsh, in *Solid State Physics*, edited by F. Seitz and D. Turnbull (Academic, New York, Vol. VI, 1958), pp. 1–63.

59F01   H. Fritzche, Phys. Rev. **115**, 336–345 (1959).

60D02   G.E. Duvall, D.E. Davenport, and J.J. Kelly, Metallurgical Effects of Explosion-Induced Shock Waves, in Research Seminar on *High Nickel Alloys for High Temperatures, Iron-Nickel Alloys, Stainless Steels* (The International Nickel Co., New York, 1960).

60K01   R.W. Keyes, in *Solid State Physics, Vol 11* edited by F. Seitz and D. Turnbull (Academic, New York, Vol. 11, 1960), pp. 149–221.

60M01   R.G. McQueen and S.P. Marsh, J. Appl. Phys. **31**, 1253–1269 (1969).

61C01   D.R. Curran, J. Appl. Phys. **32**, 1811–1814 (1961).

61D01   P.S. DeCarli and J.C. Jamieson, Science **133**, 821–822 (1961).

61D02   G.E. Duvall, in *Response of Metals to High Velocity Deformation*, edited by

202   References

        P.G. Shewmon and V.R. Zackay (Interscience, New York, 1961), pp. 165–202.
61E01   J.O. Erkman, J. Appl. Phys. **32**, 939–944 (1961).
61F01   G.R. Fowles, Stanford Research Institute, Poulter Labs. Tech. Rept. No. 003-61 (1961).
61F02   G.R. Fowles, J. Appl. Phys. **32**, 1475–1487 (1961).
61H01   H.G. Hopkins, Appl. Mech. Rev. **66**, 417–431 (1961).
61H02   K. Hruska, Czech. J. Phys. B **11**, 150–152 (1961).
61M01   S. Minshall, in *Response of Metals to High Velocity Deformation*, edited by P.G. Shewmon and V.R. Zackay (Interscience, New York, 1961), pp. 249–274.
61R01   C. Reynolds and G.E. Seay, J. Appl. Phys. **32**, 1401–1402 (1961).
61S01   P.G. Shewmon and Z.F. Zackay, Eds., *Response of Metals to High Velocity Deformation* (Interscience, New York, 1961).
62D01   G.E. Dieter, in *Strengthening Mechanisms in Solids* (American Society for Metals, Metals Park, Ott, 1962), pp. 279–340.
62D03   W.E. Deal, Jr., in *Modern Very High Pressure Techniques*, (Butterworths, London, 1962), pp. 200–227.
62G01   R.A. Graham, J. Appl. Phys. **33**, 1755–1758 (1962).
62N01   F.W. Neilson and W.B. Benedick, Sandia Corporation Report No. SCR-502 (1962).
62N02   F.W. Neilson, W.B. Benedick, W.P. Brooks, R.A. Graham and G.W. Anderson, in *Les Ondes de Detonation*, edited by G. Ribaud (Editions du Centre National de la Recherche Scientifique, Paris 1962), pp. 391–419.
62S01   K.J. Schmidt-Tiedemann, in *Proceedings of the International Conference on Physics of Semiconductors*, (The Institute of Physics and the Physical Society, London, 1962), p. 191.
62W01   J. Wackerle, J. Appl. Phys. **33**, 922–937 (1962).
63D01   D.G. Doran, in *High Pressure Measurements*, edited by A.A. Giardini and E.C. Lloyd (Butterworths, London, 1963), pp. 59–86.
63D02   G.E. Duvall and G.R. Fowles, in *High Pressure Physics and Chemistry*, edited by R.S. Bradley (Academic, New York, 1963), Vol. 2, pp. 209–291.
63D03   D.R. Curran, J. Appl. Phys. **34**, 2677–2685 (1963).
63G01   I. Goroff and L. Kleinman, Phys. Rev. **132**, 1080–1084 (1963).
63K01   L. Knopoff, in *High Pressure Physics and Chemistry*, edited by R.S. Bradley (Academic, New York, 1963), Vol. 1, pp. 227–245.
63K02   L. Knopoff, in *High Pressure Physics and Chemistry*, edited by R.S. Bradley (Academic, New York, 1963), Vol. 1, pp 247–263.
63K03   Y. Kimura, Jpn. J. Appl. Phys. **2**, 312 (1963).
63K04   J.S. Kouvel, in *Solids Under Pressure*, edited by W. Paul and D.M. Warschauer (McGraw-Hill, New York, 1963), pp. 277–301.
63M01   D.J. Milton and P.S. DeCarli, Science **140**, 670–671 (1973).
63P01   W. Paul, in *Solids Under Pressure*, edited by W. Paul and D.M. Warschauer (McGraw-Hill, New York, 1963), pp. 179–249.
63P03   W. Paul and H. Brooks, in *Progress in Semiconductors*, edited by A.F. Gibson and R.E. Burgess (Wiley, New York, 1963), Vol. 7, pp. 135–235.
63T01   J.W. Taylor, J. Appl. Phys. **34**, 2727–2731 (1963).
64C01   W.G. Cady, *Piezoelectricity* (Dover, New York, 1964), Vol. 1
64D01   L.V. Dakhovskii, Sov. Phys. Solid State, **5**, 1695–1699 (1964).

64J01   O.E. Jones and J.R. Holland, J. Appl. Phys. **35**, 1771–1773 (1964).

64M01   R.G. McQueen, in *Metallurgy at High Pressures and High Temperatures*, edited by K.A. Geschneider, Jr., M.F. Hepworth, and N.A.D. Partee. (Gordon and Breach, New York, 1964) pp. 44–132.

64P01   E.G.S. Paige, in *Progress in Semiconductors*, edited by A.F. Gibson and R.E. Burgess (Wiley, New York, 1964), Vol. 8.

64W01   M.L. Wilkins, in *Methods of Computational Physics*, edited by B. Alder, S. Fernbach, and M. Kotenberg. (Academic, New York, 1964), Vol. 3, pp. 211–263.

65A01   L.V. Al'tshuler, Sov. Phys. Usp. **8**, 52–91 (1965).

65A02   G.A. Adadurov, I.M. Barkalov, V.I. Gol'danskii, A.N. Dremin, and T.N. Ignatovich, Polym. Sci. USSR **7**, 196–197 (1965).

65A03   A.S. Appleton, Appl. Mater. Res. **4**, 195–201 (1965).

65B01   S.S. Batsanov, A.A. Deribas, E.V. Dulepov, M.G. Ermakov, and V.M. Kudinov, Combust. Explos. Shock Waves **1**, 47–49 (1965).

65B03   W.P. Brooks, J. Appl. Phys. **36**, 2788–2790 (1965).

65B04   I. Balslev, Solid State Commun. **3**, 213–218 (1965).

65G01   R.A. Graham, F.W. Neilson and R.A. Graham, J. Appl. Phys. **36**, 1775–1783 (1965).

65H01   G.E. Hauver, J. Appl. Phys. **36**, 2113–2118 (1965).

65H02   Y. Horiguchi and Y. Nomura, Carbon **2**, 436–437 (1965).

65H02   Y. Horiguchi and Y. Nomura, Chem. Ind. London, Oct., 1791–1792 (1965).

65T01   C. Truesdell and W. Noll, in *Handbuch der Physik, Band III/3*, edited by S. Flugge (Springer-Verlag, Berlin, 1965), pp. 1–602.

66A04   M. Asche, O.G. Sarbej, and V.M. Vasetskii, Phys. Status Solid **18**, 749–754 (1966).

66B01   O.R. Bergmann and J. Barrington, J. Am. Ceram. Soc. **49**, 502–507 (1966).

66B02   I. Baslev, Phys. Rev. **143**, 636–647 (1966).

66D01   D.G. Doran and R.K. Linde, in *Solid State Physics*, edited by F. Seitz and D. Turnbull (Academic, New York, 1966), Vol. 19, pp. 229–290.

66G01   R.A. Graham, O.E. Jones, and J.R. Holland, J. Phys. Chem. Solids **27**, 1519–1529 (1965).

66H01   J.A. Hedvall, *Solid State Chemistry: Whence, Where and Whither*, (Elsevier, New York, 1966).

66H02   W.J. Halpin, J. Appl. Phys. **37**, 153–163 (1966).

66R01   S. Riskaer, Phys. Rev. **152**, 845–849 (1966).

67C01   E.C.T. Chao, Science, **156**, 192–202 (1967).

66Z01   E.G. Zukas, Met. Engin. Quart. **6**, 1–20 (1966).

67G01   R.A. Graham, D.H. Anderson, and J.R. Holland, J. Appl. Phys. **38**, 223–229 (1967).

67K02   J.D. Kennedy and W.B. Benedick, Solid State Commun. **5**, 53–55 (1967).

67O01   H.E. Otto and R. Mikesell, in *Proceedings of the First International Conference of the Center for High Energy Rate Forming* (Denver Research Institute, University of Denver, Denver, co, 1967, vol. 2) pp. 7.6.1–7.6.46.

67S02   O.M. Stuetzer, J. Appl. Phys. **38**, 3901–3904 (1967).

67S03   O.M. Stuetzer, J. Acoust. Soc. Am. **42**, 502–508 (1967).

68B04   K. Bulthius, Phillips Res. Rep. **23**, 25–47 (1968).

68C01   G.R. Cowan, B.W. Dunnington, and A.H. Holtzman, Process for Synthesizing Diamond, U.S. Patent No. 3,401,019, September 10, 1968.

68D01   A.N. Dremin and O.N. Breusov, Russ. Chem. Rev. **37**, 392–402 (1968).

68D02   D.G. Doran, J. Appl. Phys. **39**, 40–47 (1968).

68G03   R.A. Graham, J. Appl. Phys. **39**, 437–439 (1968).

68G04   R.A. Graham and W.J. Halpin, J. Appl. Phys. **39**, 5077–5082 (1968).

68G05   R.A. Graham and G.E. Ingram, in *Behavior of Dense Media Under High Dynamic Pressures*, edited by J. Berger (Gordon and Breach, New York, 1968), pp. 469–482.

68K02   S.B. Kormer, Sov. Phys. Usp. **11**, 229–254 (1968).

68K03   J.D. Kennedy in *Behavior of Dense Media Under High Dynamic Pressures*, edited by J. Berger (Gordon and Breach, New York, 1968), pp. 407–418.

68P02   F.H. Pollak and M. Cardona, Phys. Rev. **172**, 816–837 (1968).

69A01   L.V. Al'tshuler and A.A. Bakanova, Sov. Phys. Usp. **11**, 678–689 (1969).

69A02   T.J. Ahrens, D.L. Anderson, and A.E. Ringwood, Rev. Geophys. **7**, 667–707 (1969).

69G01   Group GMX-6, Los Alamos Scientific Laboratory Report No. LA-4167-MS (1969).

69H01   W. Herrmann, in *Wave Propagation in Solids*, edited by J. Miklowitz (American Society of Mechanical Engineers, New York, 1969), pp. 129–183.

69H02   W. Herrmann, J. Appl. Phys. **40**, 2490–2499 (1969).

69K01   H. Kawai, Jpn. J. Appl. Phys. **8**, 975 (1969).

69S02   J.F. Schetzina and J.P. McKelvey, Phys. Rev. **181**, 1191–1195 (1969).

69S03   L.R. Saravia and D. Brust, Phys. Rev. **178**, 1240–1243 (1969).

69T01   R.N. Thurston, J. Acoust. Soc. **45**, 1329–1341 (1969).

69W01   R.C. Wayne, J. Appl. Phys. **40**, 15–22 (1969).

70B01   L.M. Barker and R.E. Hollenbach, J. Appl. Phys. **41**, 4208–4226 (1970).

70B02   R.R. Boade, Experimental Shock Loading Properties of Porous Materials and Analytical Methods to Describe these Properties, Sandia Laboratories Report No. SC-DC-70-5052, August, 1970.

70C01   B. Crossland and J.D. Williams, Metall. Rev. **144**, 79–100 (1970).

70H02   G.E. Hauver, in *Fifth Symposium (International) on Detonation*, edited by S.J. Jacobs and R. Roberts (U.S. GPO, Washington, DC, 1970), pp. 387–397.

70I01   G.E. Ingram and R.A. Graham, in *Fifth Symposium (International) on Detonation*, edited by S.J. Jacobs and R. Roberts (U. S. GPO, Washington, DC, 1970), pp. 369–386.

70M01   R.G. McQueen, S.P. Marsh, J.W. Taylor, J.N. Fritz and W.J. Carter, in *High Velocity Impact Phenomena*, edited by R. Kinslow (Academic, New York, 1970), pp. 293–417, 515–568.

70M03   P.J. Melz, Phys. Chem. Solids **32**, 209–221 (1970).

70S01   D.L. Styris and G.E. Duvall, High Temp.-High Pressures **2**, 447–499 (1970).

71D01   L. Davison and A.L. Stevens, Sandia Labortories Report No. SC-TM-70-786 (1971).

71G01   R.A. Graham and W.P. Brooks, J. Phys. Chem. Solids **32**, 2311–2330 (1971).

71G06   W.H. Gust and E.B. Royce, J. Appl. Phys. **42**, 1897–1905 (1971).

71J01   J.N. Johnson, J. Appl. Phys. **42**, 5522–5530 (1971).

71J02   O.E. Jones and R. A. Graham, in *Accurate Characterization of the High Pressure Environment*, edited by E.C. Lloyd (U. S. GPO, Washington, DC, 1971) pp. 229–241.

71K01   R.N. Keeler, in *Proceedings of the International School of Physics "Enrico*

Fermi," *Course XLVII, Physics of High Energy Density*, edited by P. Caldirola and H. Knoepfel (Academic, New York, 1971), pp. 51–80.

71K02   R.N. Keeler, in *Proceedings of the International School of Physics "Enrico Fermi," Course XLVII, Physics of High Energy Density*, edited by P. Caldirola and H. Knoepfel (Academic, New York, 1971), pp. 106–126.

71R01   E.B. Royce, in *Proceedings of the International School of Physics "Enrico Fermi," Course XLVII, Physics of High Energy Density*, edited by P. Caldirola and H. Knoepfel (Academic, New York, 1971), pp. 80–95.

71R02   E.B. Royce, in *Proceedings of the International School of Physics "Enrico Fermi," Course XLVII, Physics of High Energy Density*, edited by P. Caldirola and H. Knoepfel (Academic, New York, 1971), pp. 95–106.

71R03   E.B. Royce, *Proceedings of the International School of Physics "Enrico Fermi," Course XLVII, Physics of High Energy Density*, edited by P. Caldirola and H. Knoepfel (Academic, New York, 1971), pp. 126–138.

71T01   R.F. Trunin, G.V. Simakov, and M.A. Podurets, Acad. Sci. USSR, Bulletin, Phys. Solid Earth, No. 2, 102–106 (1971).

72A01   T.J. Ahrens, Tectonophys. **13**, 189–219 (1972).

72A02   J.R. Asay, G.R. Fowles, G.E. Duvall, M.H. Miles, and R.F. Tinder, J. Appl. Phys. **43**, 2132–2145 (1972).

72A05   G. Arlt, in *Proceedings of the International School of Physics, Atomic Structure, and Properties of Solids*, edited by E. Burstein (Academic, New York, 1972), pp. 201–205 (1972).

72C04   N.L. Coleburn, J.W. Forbes, and H.D. Jones, J. Appl. Phys. **43**, 5007–5012 (1972).

72C05   M.M. Carroll and A.C. Holt, J. Appl. Phys. **43**, 1626–1635 (1972).

72G01   R.A. Graham, Phys. Rev. **6**, 4779–4792 (1972).

72G02   R.A. Graham, J. Acoust. Soc. Am. **52**, 1578–1581 (1972).

72G03   R.A. Graham and G.E. Ingram, J. Appl. Phys. **43**, 826–835 (1972).

72G05   W.H. Gust and E.B. Royce, J. Appl. Phys. **43**, 4437–4442 (1972).

72H01   R.R. Hasiguti, in *Annual Review of Material Science*, edited by R.A. Huggins, R.H. Bube, and R.W. Roberts (Annual Reviews, Palo Alto, 1972), Vol. 2, pp. 69–92.

72J01   O.E. Jones, in *Behavior and Utilization of Explosives in Engineering Design*, edited by L. Davison, J.E. Kennedy, and F. Coffey (New Mexico Section, American Society of Mechanical Engineers, New York, 1972), pp. 125–148.

72L02   P.C. Lysne, J. Appl. Phys. **43**, 425–431 (1972).

72M01   R.M. Martin, Phys. Rev. **5**, 1607–1613 (1972).

72M02   R.M. Martin, in *Proceedings of the International School of Physics, Atomic Structure, and Properties of Solids*, edited by E. Burstein (Academic New York, 1972), pp. 492–500.

72S04   D. Stöffler, Fortschr. Miner. **49**, 50–113 (1972).

73A01   G.A. Adadurov, V.I. Gol'danskii, and P.A. Yampol'skii, Mendeleev Chem. J **18**, 92–103 (1973).

73B01   A.A. Bakanova, I.P. Dudoladov, and Yu. N. Sutulov, J. Appl. Mech. Appl. Phys. No. 2, 241–245 (1973).

73D01   G.E. Duvall, in *Dynamic Response of Materials to Intense Impulsive Loading*, edited by P.C. Chou and A.K. Hopkins (Wright-Patterson AFB, Ohio, 1973), pp. 89–121.

73F01   G.R. Fowles, in *Dynamic Response of Materials to Intense Impulsive Loading*, edited by P.C. Chou and A.K. Hopkins (Wright-Patterson AFB, Ohio, 1973), pp. 405–480.

73G01   W.H. Gust, A.C. Holt, and E.B. Royce, J. Appl. Phys. **44**, 550–560 (1973).

73H01   W. Herrmann and J.W. Nunziato, in *Dynamic Response of Materials to Intense Impulsive Loading*, edited by P.C. Chou and A.K. Hopkins (Wright-Patterson AFB, Ohio, 1973), pp. 123–281.

73H02   W. Herrmann and D.L. Hicks, in *Metallurgical Effects at High Strain Rates*, edited by R.W. Rohde, B.M. Butcher, J.R. Holland, and C.H. Karnes (Plenum, New York, 1973), pp. 57–91.

73J03   C.L. Julian and F.O. Lane, Jr., Phys. Rev. B **7**, 723–728 (1973).

73L01   W.C. Leslie, in *Metallurgical Effects at High Strain Rates*, edited by R.W. Rohde, B.M. Butcher, J.R. Holland, and C.H. Karnes (Plenum, New York, 1973), pp. 571–586.

73L02   C.Y. Liu, K. Ishizaki, J. Pasauwe, and I.L. Spain, High Temp.-High Pressures **5**, 359–388 (1973).

74E01   L.R. Edwards and L.C. Bartel, Phys. Rev. B **10**, 2044–2048 (1974).

74G01   R.A. Graham, J. Phys. Chem. Solids **35**, 355–372 (1974).

74H02   R.C. Hanson, K. Helliwell, and C. Schwab, Phys. Rev. B **9**, 2649–2614 (1974)

74H03   W.A. Harrison, Phys. Rev. B **10**, 767–770 (1974).

74H04   D.B. Hayes, J. Appl. Phys. **45**, 1208–1217 (1974).

74M01   W.J. Murri, D.R. Curran, C.F. Peterson, and R.C. Crewdson, in *Advances in High Pressure Research*, edited by R.H. Wentorf, Jr. (Academic, New York, 1974), Vol. 4, pp. 1–163.

74N01   J.W. Nunziato, E.K. Walsh, K.W. Schuler, and L.M. Barker, in *Handbuch der Physic, Band VIa/4*, edited by S. Flugge (Springer-Verlag, Berlin, 1974), pp. 1–108.

74S01   K.W. Schuler and J.W. Nunziato, Rheol. Acta. **13**, 265–273 (1974).

74T01   R.N. Thurston, in *Handbuch der Physic, Band VIa/4*, edited by S. Flugge (Springer-Verlag, Berlin, 1974), pp. 109–308.

74T02   R.F. Trunin, G.V. Simakov, and M.A. Podurets, Acad. Sci. USSR, Bull. Earth Phys. **12**, 789–792 (1974).

75A01   J.R. Asay and D.B. Hayes, J. Appl. Phys. **46**, 4789–4800 (1975).

75D01   J.J. Dick and D.L. Styris, J. Appl Phys. **46**, 1602–1617 (1975).

75G04   R.A. Graham, J. Appl. Phys. **46**, 1901–1909 (1975).

75G05   R.A. Graham, and P.C. Chen, Solid State Commun. **17**, 469–471 (1975).

75G06   R.A. Graham and L.C. Yang, J. Appl. Phys. **46**, 5300–5301 (1974).

75K02   A.I. Korobov and L.E. Lyamov, Sov. Phys. Solid State **17**, 932–933 (1975)

76B01   A.S. Bedford, D.S. Drumheller and H.J. Sutherland, in *Mechanics Today*, edited by S. Nemat-Nasser (Pergamon, New York, 1976), Vol. 3, pp. 1–54.

76B02   F. Bauer and K. Vollrath, Ferroelectrics **12**, 153–156 (1976).

76C02   P.J. Chen, L. Davison, and M.F. McCarthy, J. Appl. Phys. **47**, 4759–4764 (1976).

76G04   R.A. Graham, Ferroelectrics **10**, 65–69 (1976).

76H01   W. Herrmann, in *Propagation of Shock Waves in Solids*, edited by E. Varley (American Society of Mechanical Engineers, New York, 1976), pp. 1–26.

76M01   V.N. Mineev and A.G. Ivanov, Sov. Phys. Usp. **19**, 400–419 (1976).

77D01   G.E. Duvall and R.A. Graham, Rev. Modern Phys. **49**, 523–579 (1977).

77D02   L. Davison, A.L. Stevens, and M.E. Kipp, J. Mech. Phys. Solids **25**, 11–28 (1972).

77G01   D.E. Grady, in *High Pressure Research Applications in Geophysics*, edited by M.H. Manghnani and S. Akimoto (Academic, New York, 1977), pp. 389–438.

77G05   R.A. Graham, IEEE Trans. Sonics Ultrason. **SU-24**, 137 (1977)

77G06   R.A. Graham, J. Appl. Phys. **48**, 2153–2163 (1977).

77G07   D.E. Grady, in *High Pressure Research: Applications in Geophysics*, edited by M.H. Manghnani and S. Akimoto (Academic, New York, 1977) pp. 389–438.

77H01   D.B. Hayes, Sandia Laboratories Report No. SAND77-0267C, 1977.

77H04   K. Helliwell, R.C. Hanson, and C. Schwab, in *Proceedings of the 1977 IEEE Ultrasonics Symposium* (IEEE, New York, 1977). 317–320.

77L01   R.J. Lawrence and L. Davison, in *Proc. of the Symposium on Applied Computer Methods in Engineering*, edited by L.C. Welford, (School of Engineering, University of Southern California, Los Angeles, 1977), Vol. II, pp. 941–950.

77L02   J. Lipkin and J.R. Asay, J. Appl. Phys. **48**, 182–189 (1977).

77N02   E.Z. Novitskii, V.A. Ogorodnikov, and S.Ya. Pinchuk, Combust., Explos. Shock Waves (USSR) **13**, 221–234 (1977).

77S01   P.L. Stanton and R.A. Graham, Appl. Phys. Lett. **31**, 723–725 (1977).

77T02   Y.F. Tsay and B. Bendow, Phys. Rev. B **16**, 2663–2675 (1977).

77V01   M. Van Thiel, J. Shaner, and E. Salinas, Lawrence Livermore Laboratory Report No. UCRL 50108, Vols. 1 and 2, Rev. 1, Vol. 3 (1977).

78A01   L.V. Al'tshuler, J. Appl. Mech. Tech. Phys. **19**, 496–505 (1978).

78G01   R.A. Graham and J.R. Asay, High Temp.-High Pressures **10**, 355–390 (1978).

78G02   R.A. Graham and R.P. Reed, Eds., Selected Papers on Piezoelectricity and Impulsive "Pressure" Measurement, Sandia Laboratories Report No. SAND78-1911, 1978, third reprinting 1989.

78H03   C.K. Hruska, IEEE Trans. Sonics Ultrason. **SU-25**, 198–203 (1978).

78K01   R.G. Kepler, Ann. Rev. Phys. Chem. **29**, 497 (1978).

78L05   P.C. Lysne, J. Appl. Phys. **49**, 4296–4297 (1978).

78Y01   V.V. Yakushev, Combust. Explos. Shock Waves, (USSR) **14**, 131–146 (1978).

79D01   L.Davison and R.A. Graham, Phys. Rep. **55**, 256–379 (1979).

79G01   R.A. Graham, J. Phys. Chem. **83**, 3048–3056 (1979).

79G02   R.A. Graham, in *High Pressure Technology, Sixth AIRAPT Conference, Vol. 2, Applications and Mechanical Properties*, edited by K.D. Timmerhaus and M.S. Barber (Plenum, New York, 1979), pp. 854–869.

79H01   W.F. Hemsing, Rev. Sci. Instrum. **50**, 73–79 (1979).

79N03   E.Z. Novitskii, V.C. Sadunov and G.Ya. Karpenko, Combust. Explos. Shock Waves **14**, 505–516 (1979).

79P03   P.O. Pashkov, S.P. Pisarev, V.D. Rogozin and A.F. Trudov, in *Physics of Shock Pressures: Proceedings Second All-Union Symposium on Shock Pressures, 1976*, edited by S.S. Batsanov (All-Union Scientific Research Insitute of Physical Engineering and Radiotechnical Measurements, Moscow, 1979), translation, Sandia Laboratories Report No. SAND81-6006, February, 1981.

80A01   T.J. Ahrens, Science **207**, 1035–1041 (1980).

80C01   L.L. Chhabildas and J.W. Swegle, J. Appl. Phys. **51**, 4799–4807 (1980).

80G01   R.A. Graham, Bull. Am. Phys. Soc. **25**, 495 (1980).

80G02   R.A. Graham, in *Megagauss Physics and Technology*, edited by P.J. Turchi (Plenum, New York, 1980), pp. 147–150.

80G03   W.H. Gust, in *High Pressure Science and Technology*, edited by B. Vodar and Ph. Marteau (Pergamon, New York, 1980), vol. 2, pp. 1009–1018.

80G04   Y.M. Gupta, D.D. Keough, D.F. Walter, K.C. Dao, D. Henley, and A. Urweider, Rev. Sci. Instrum. **51**, 183–194 (1980).

80M01   S.P. Marsh, Ed., *LASL Shock Hugoniot Data* (University of California, Berkeley, 1980).

80N01   W.J. Nellis and A.C. Mitchell, J. Chem. Phys. **73**, 6137–6145 (1980).

80T01   K. Tashiro, M. Kobayashi, H. Tadokoro, and E. Fukada, Macromolecules **13**, 691–698 (1980).

81A01   G.A. Adadurov and V.I. Gol'danskii, Russ. Chem. Rev. **50**, 1810–1827 (1981).

81G01   D.E. Grady, Appl. Phys. Lett. **38**, 825–826 (1981).

81G02   R.A. Graham, in *Shock Waves and High-Strain-Rate Phenomena in Metals*, edited by M.A. Meyers and L.E. Murr (Plenum, New York, 1981), pp. 375–386.

81K01   B.A. Khasainov, A.A. Borisov, B.S. Ermolaev and A.I. Korotkov, in *Proceedings, Second All-Union Conference on Detonation, No. 2*, edited by A.N. Dremin October 20–22 1981 (Institute of Chemical Physics, Chernogolovka), pp. 19–22.

81K02   B.A. Khasainov, A.A. Borisov, B.S. Ermolaev, and A.I. Korotkov, in *Proceedings, Seventh Symposium (International) on Detonation*, edited by J. Short, Naval Surface Weapons Center Report No. NSWC MP82-334, 1981, pp. 435–447.

81N01   S.A. Novikov, J. Appl. Mech. Phys. **22**, 385–394 (1981).

81S01   H. Schmalzried, *Solid State Reactions*, (Verlag Chemie, Basel 1981).

82B01   F. Bauer, in *Shock Waves in Condensed Matter—1981*, edited by W.J. Nellis et al. (American Institute of Physics, New York, 1982), pp. 251–267.

82G01   J. Golden, M. Sc. Thesis, University of New Mexico, 1982.

82G02   R.A. Graham, in *Shock Waves in Condensed Matter—1981*, edited by W.J. Nellis et al. (American Institute of Physics, New York, 1982), pp. 52–56.

82H01   D.L. Hankey, R.A. Graham, W.F. Hammetter, and B. Morosin, J. Mater. Sci. Lett. **1**, 445–447 (1982).

83G01   Y.M. Gupta, J. Appl. Phys. **54**, 6256–6266 (1983).

83G02   R.A. Graham, B. Morosin, and B. Dodson, The Chemistry of Shock Compression: A Bibliography, Sandia National Laboratories Report No. SAND83-1887, October, 1983.

83M01   M.A. Meyers and C.T. Aimone, *Prog. Mater. Sci.*, 1–96 (1983).

83N01   W.J. Nellis, M. Ross, A.C. Mitchell, M. van Thiel, D.A. Young, J.H. Ree, and R.J. Trainor, Phys. Rev. A **27**, 608–611 (1983).

83S01   P.L. Stanton and R.A. Graham, Ferroelectrics **49**, 177–182 (1983).

84B01   E.K. Beauchamp, R.E. Loehman, R.A. Graham, B. Morosin and E.L. Venturini, in *Emergent Process Methods For High-Technology Ceramics*, edited by R.F. Davis, H. Palmour, III and R.F. Davis (Plenum, New York, 1984), Vol. 17, pp. 735–748.

84B02   J.M. Brown and J.W. Shaner, in *Shock Waves in Condensed Matter-1983*, edited by J.R. Asay, R.A. Graham, and G.K. Straub (North-Holland, Amsterdam, 1984), pp. 91–94.

84G01   R.A. Graham and D.M. Webb, in *Shock Waves in Condensed Matter—1983*, edited by J.R. Asay, R.A. Graham, and G.K. Straub (North Holland, Amsterdam, 1984), pp. 211–214.

84G04   Y.M. Gupta, Theoretical and Experimental Studies to Develop a Piezoelectric Shear Stress Interface Gage, SRI International Report, 1984.

84H01   W.F. Hammetter, J.R. Hellman, R.A. Graham, and B. Morosin, in *Shock Waves in Condensed Matter-1983*, edited by J.R. Asay, R.A. Graham, and G.K. Straub (North-Holland, Amsterdam, 1984), pp. 391–394.

84H02   J.R. Hellmann, K. Kuroda, A.H. Heuer, and R.A. Graham, in *Shock Waves in Condensed Matter-1983*, edited by J.R. Asay, R.A. Graham, and G.K. Straub (North-Holland, Amsterdam, 1984), pp. 387–390.

84M01   B. Morosin and R.A. Graham, Mater. Sci. Engin. **66**, 73–87 (1984).

84M02   R.G. McQueen, J.N. Fritz, and C. E. Morris, in *Shock Waves in Condensed Matter-1983*, edited by J.R. Asay, R.A. Graham, and G.K. Straub (North-Holland, Amsterdam, 1984), pp. 95–98.

84M03   A.K. McMahan, Phys. Rev. B **29**, 5982–5985 (1984).

84S01   R.B. Schwarz, P. Kasiraj, T. Vreeland, and T.J. Ahrens, Acta Metall. **32**, 1243–1252 (1984).

84S02   Y. Syono, in *Materials Science of the Earth's Interior*, edited by I. Sunagawa (Terra, Tokyo, 1984), pp. 395–414.

84T01   J. Taylor in *Shock Waves in Condensed Matter-1983*, edited by J.R. Asay, R.A. Graham, and G.K. Straub (North-Holland, Amsterdam, 1984), pp. 1–15.

84V01   E.L. Venturini and R.A. Graham, in *Defect Properties and Processing of High Technology Nonmetallic Materials*, edited by J.H. Crawford, Jr. Y. Chen and W.A. Sibley (North Holland, Amsterdam, 1984), pp. 383–389.

84W01   A.R. West, *Solid State Chemistry and its Applications* (Wiley, New York, 1984).

85B01   E.K. Beauchamp, M.J. Carr, and R.A. Graham, J. Am. Ceram. Soc. **68**, 696–699 (1985).

85H01   Y. Horie, R.A. Graham, and I.K. Simonsen, Mater. Lett. **3**, 354–359 (1985).

85K01   K. Kondo, S. Soga, A. Sawaoka, and M. Araki, J. Mater. Sci. **20**, 1033–1048 (1985).

85L01   Y.K. Lee, F.L. Williams, R.A. Graham, and B. Morosin, J. Mater. Sci. **20**, 2488–2496 (1985).

85S01   J.W. Swegle and D.E. Grady, J. Appl. Phys. **58**, 692–701 (1985).

86B01   F. Bauer, U.S. Patent, No. 4,611,260, Method and Device for Polarizing Ferroelectric Materials, September 9, 1986.

86C01   M.J. Carr and R.A. Graham, in *Shock Waves in Condensed Matter*, edited by Y.M. Gupta (Plenum, New York, 1986), pp. 803–808.

86D01   G.E. Duvall, in *Metallurgical Applications of Shock Wave and High-Strain-Rate Phenomena*, edited by L. Murr, K.P. Staudhammer and M.A. Meyers, (Marcel Dekker, New York, 1986), pp. 3–25.

86G01   R.A. Graham, B. Morosin, Y. Horie, E.L. Venturini, M. Boslough, M.J. Carr, and D.L. Williamson, in *Shock Waves in Condensed Matter*, edited by Y.M. Gupta (Plenum, New York, 1986), pp. 693–711.

86G02   R.A. Graham, B. Morosin, E.L. Venturini, and M.J. Carr, Annu. Rev. Mater. Sci. **16**, 315–341 (1986).

86G03   R.A. Graham and M.J. Carr, in *Shock Waves in Condensed Matter*, edited by Y.M. Gupta (Plenum, New York, 1986), pp. 803–808.

86G04   R.A. Graham and D.M. Webb, in *Shock Waves in Condensed Matter*, edited by Y.M. Gupta (Plenum, New York, 1986), pp. 831–836.

86G05   W.H. Gourdin, Prog. Mater. Sci. **30**, 39 (1986).

86H01   Y. Horie, D.E.P. Hoy, I. Simonsen, R.A. Graham, and B. Morosin, in *Shock Waves in Condensed Matter*, edited by Y.M. Gupta (Plenum, New York, 1986), pp. 749–754.

86H02   Y. Horie, R.A. Graham, and I.K. Simonsen, in *Metallurgical Applications of Shock Wave and High-Strain-Rate Phenomena*, edited by L. Murr, K.P. Staudhammer, and M.A. Meyers (Marcel Dekker, New York 1986), pp. 1023–1035.

86M01   B. Morosin, E.L. Venturini, and R.A. Graham, in *Shock Waves in Condensed Matter*, edited by Y.M. Gupta (Plenum, New York, 1986), pp. 797–801.

86M02   B. Morosin and R.A. Graham, in *Metallurgical Applications of Shock Wave and High-Strain-Rate Phenomena*, edited by L. Murr, K.P. Staudhammer, and M.A. Meyers (Marcel Dekker, New York, 1986), pp. 1037–1047.

86V01   E.L. Venturini, B. Morosin, and R.A. Graham, in *Shock Waves in Condensed Matter*, edited by Y.M. Gupta (Plenum, New York, 1986), pp. 815–820.

86W01   J.L. Wise and L.C. Chhabildas, in *Shock Waves in Condensed Matter*, edited by Y.M. Gupta (Plenum, New York, 1986), pp. 441–454.

86W02   D.L. Williamson, B. Morosin, E.L. Venturini, and R.A. Graham, in *Shock Waves in Condensed Matter*, edited by Y.M. Gupta (Plenum, New York, 1986), pp. 809–814.

86W03   D.C. Williamson, E.L. Venturini, R.A. Graham, and B. Morosin, Phys. Rev. B **34**, 1899–1907 (1986).

86W04   F.L. Williams, Y.K. Lee, B. Morosin, and R.A. Graham, in *Shock Waves in Condensed Matter*, edited by Y.M. Gupta (Plenum, New York, 1986), pp. 791–796.

87A01   T.J. Ahrens, Methods Experimen. Phys. **24**, 185–235 (1987).

87A02   A.V. Attetkov and V.S. Solovev, Combust. Explos. Shock Waves (USSR) **23**, 482–491 (1987).

87A03   T.J. Ahrens and J.D. O'Keefe, Int. J. Impact Eng. **5**, 13–32 (1987).

87A04   J.R. Asay and G.I. Kerley, Int. J. Impact Eng. **5**, 69–99 (1987).

87B01   F. Bauer, U.S. Patent No. 4,684,337, Device for Polarizing Ferroelectric Materials, August 4, 1987.

87B02   E.K. Beauchamp and M.J. Carr, in *High Pressure Explosive Processing of Ceramics*, edited by R.A. Graham and A.B. Sawaoka (Transtech, 1987), pp. 175–206.

87B03   E.K. Beauchamp, R.A. Graham, and M.J. Carr, *Material Research Society Symposium Proceeding*, edited by J.H. Crawford, Jr., *et al.* (North-Holland, Amsterdam, 1984), Vol. 24, pp. 281–289.

87B04   E.K. Beauchamp, M.J. Carr, and R.A. Graham, Adv. Ceram Mater. **2**, 79–84 (1987).·

87C01   L.C. Chhabildas and R.A. Graham, in *Techniques and Theory of Stress Measurements for Shock Wave Applications*, edited by R.R. Stout, E.R. Norwood, and M.E. Fourney. (American Society of Mechanical Engineers, New York, 1987) AMD-Vol. 83, pp. 1–18.

87C02   J. Cagnoux, P. Chartagnac, P. Hereil, and M. Perez, Ann. Phys. **12**, 451–524 (1987).

87C03   M.J. Carr in *High Pressure Explosive Processing of Ceramics*, edited by R.A. Graham and A.B. Sawaoka (Transtech, 1987), pp. 341–376.

87C04   D.R. Curran, L. Seaman and D.A. Shockey, Phys. Rep. **147**, 253–288 (1987).

87E01   V.E. Egorushkin, V.E. Panin, E.V. Cavushkin, and Yu.A. Khon, Sov. Phys. J. **1**, 5–24 (1987).

87G01   Y.M. Gupta and S.C. Gupta, J. Appl. Phys. **61**, 489–498 (1987).

87G02   R.A. Graham and A.B. Sawaoka, Eds., *High Pressure Explosive Processing of Ceramics* (Transtech, Aedermannsdorf, Switzerland, 1987).

87G03   R.A. Graham, in *High Pressure Explosive Processing of Ceramics*, edited by R.A. Graham and A.B. Sawaoka (Transtech, Aedermannsdorf, Switzerland, 1987), pp. 30–64.

87J01   W.E. Johnson and C.E. Anderson, Jr., Int. J. Impact Eng. **5**, 423–439 (1987).

87M01   B. Morosin, in *High Pressure Explosive Processing of Ceramics*, edited by R.A. Graham and A.B. Sawaoka (Transtech, 1987), pp. 285–339.

87O01   A.I. Olemskoi and V.A. Petrunin, Sov. Phy. J. **1**, 61–92 (1987).

87V01   E.L. Venturini, B. Morosin, R.A. Graham, D.L. Williamson, and F.L. Williams, Sandia National Laboratories Report No. SAND87-0315, June, 1987.

88B01   B.L. Holian, Phys. Rev. A **37**, 2562 (1988).

88C01   W.H. Casey, M.J. Carr, and R.A. Graham, Geochim. Cosmochim. Acta **52**, 1545–1556 (1980).

88C02   M.P. Conner, M.Sc. Thesis, Washington State University, Pullman, WA, 1988.

88C03   L.C. Chhabildas, in *Shock Waves in Condensed Matter 1987*, edited by S.C. Schmidt, J.N. Johnson, and L.W. Davison (North-Holland, Amsterdam, 1987), pp. 579–584.

88C04   L.C. Chhabildas and L.M. Barker, in *Shock Waves in Condensed Matter 1987*, edited by S.C. Schmidt, J.N. Johnson, and L.W. Davison (North-Holland, Amsterdam, 1987), pp. 111–114.

88D01   S.Yu. Korostelev, S.G. Pskh'e, and V.E. Panin, Combust. Explos. Shock Waves (USSR) **24**, 755–758 (1988).

88G01   R.A. Graham, in *Shock Waves in Condensed Matter 1987*, edited by S.C. Schmidt, J.N. Johnson, and L.W. Davison (North-Holland, Amsterdam, 1987), pp. 11–18.

88G02   R.A. Graham, B. Morosin, and D.M. Bush, in *Shock Waves in Condensed Matter 1987*, edited by S.C. Schmidt, J.N. Johnson, and L.W. Davison (North-Holland, Amsterdam, 1987), pp. 179–182.

88H01   W.F. Hammetter, R.A. Graham, B. Morosin, and Y. Horie, in *Shock Waves in Condensed Matter 1987*, edited by S.C. Schmidt, J.N. Johnson, and L.W. Davison (North-Holland, Amsterdam, 1987), pp. 431–434.

88M01   M.A. Meyers, N.N. Thadhani, and L.H. Yu, in *Shock Waves for Industrial Applications*, edited by L.E. Murr (Noyes, Park Ridge, N.J. 1988), pp. 265–335.

88S01   R.E. Setchell, in *Shock Waves in Condensed Matter 1987*, edited by S.C. Schmidt, J.N. Johnson, and L.W. Davison (North Holland, Amsterdam, 1988), pp. 623–626.

88S02   G.A. Samara and F. Bauer, in *Shock Waves in Condensed Matter 1987*, edited by S.C. Schmidt, J.N. Johnson, and L.W. Davison (North-Holland, Amsterdam, 1987), pp. 611–614.

88S03   Y. Syono, in *Shock Waves in Condensed Matter 1987*, edited by S.C. Schmidt, J.N. Johnson, and L.W. Davison (North-Holland, Amsterdam, 1987), pp. 19–26.

88T01   N.N. Thadhani, Adv. Mater. Manuf. Processes **3**, 493–549 (1988).

88T02   Z.P. Tang and Y.M. Gupta, J. Appl. Phys. **64**, 1827–1837 (1988).

88V01   E.L. Venturini, R.A. Graham, and B. Morosin, in *Shock Waves in Condensed Matter 1987*, edited by S.C. Schmidt, J.N. Johnson, and L.W. Davison (North-Holland, Amsterdam, 1987), pp. 451–454.

88W01   S.J. Work, L.H. Yu, N.N. Thadhani, M.A. Meyers, R.A. Graham, and W.F. Hammetter, in *Combustion and Plasma Synthesis of High Temperature Materials.*, edited by Z.A. Munir and J.B. Holt (VCH, New York, 1990), pp. 133–143.

89G01   R.A. Graham, in *Third International Symposium on High Dynamic Pressure*, edited by R. Cheret (Assoc. Francaise de Pyrotechnie, Paris, 1989), pp. 175–180.

89P01   V.E. Panin and Yu.V. Grinyaev, in *High Energy Rate Fabrication*, September 18–22, 1989 (Ljubljana, Yugoslavia, 1989).

89S01   S.A. Sheffield, R. Engelke, and R.R. Alcon, in *Ninth Symposium (International) on Detonation*, edited by J.M. Short (1989).

89S02   G.A. Samara, J. Polym. Sci.: Pt. B: Polym. Phys. **27**, 39–51 (1989).

89T01   N.N. Thadhani, M.J. Costello, I. Song, S. Work, and R.A. Graham, in *Solid State Powder Processing*, edited by A.H. Clauer and J.J. de Barbadillo (TMS, Warrendale, PA, 1990), pp. 97–109.

89T02   R.F. Trunin, G.V. Simakov, Yu.N. Sutulov, A.B. Medvedev, B.D. Rogozkin, and Yu.E. Fedorov, Sov. Phys. JETP **69**, 580–588 (1989).

89Z01   Y. Zhang, J.M. Stewart, B. Morosin, R.A. Graham, and C.R. Hubbard, Appl. Phys. Commun. **9**, 183–202 (1989).

90A01   M.U. Anderson, R.A. Graham, and D.E. Wackerbarth, in *Shock Compression of Condensed Matter-1989*, edited by S.C. Schmidt, J.N. Johnson, and L.W. Davison (North-Holland, Amsterdam, 1990), pp. 805–811.

90B01   E.K. Beauchamp and M.J. Carr, J. Am. Ceram. Soc. **73**, 49–53 (1990).

90B02   C.H. Bundy, R.A. Graham, S.F. Kuehn, R.R. Precit, and M.S. Rogers, U.S. Patent 4,893,049, January 9, 1990.

90B03   F. Bauer and R.A. Graham, in *Shock Compression of Condensed Matter-1989*, edited by S.C. Schmidt, J.N. Johnson, and L.W. Davison (North-Holland, Amsterdam, 1990), pp. 793–796.

90D01   A.N. Dremin and A.M. Molodets, in *Shock Compression of Condensed Matter-1989*, edited by S.C. Schmidt, J.N. Johnson, and L.W. Davison (North-Holland, Amsterdam, 1990), pp. 415–420.

90D02   J.E. Dunn, in *Shock Compression of Condensed Matter-1989*, edited by S.C. Schmidt, J.N. Johnson, and L.W. Davison (North-Holland, Amsterdam, 1990), pp. 21–32.

90E01   L.R. Edwards, private communication (1990).

90G01   G.T. Gray, III, in *Shock Compression of Condensed Matter—1989*, edited by S.C. Schmidt, J.N. Johnson, and L.W. Davison (North-Holland, Amsterdam, 1990), pp. 407–414.

90K01   H. Kunishige, Y. Oya, Y. Kukuyama, S. Watanabe, H. Tamura, A.B. Sawaoka, T. Taniguchi, and Y. Horie, Report of the Research Laboratory of Engineering Materials, Tokyo Institute of Technology, No. 15, pp. 235–264 (1990).

90O01   K. Oh and P. Persson, in *Shock Compression of Condensed matter-1989*, edited by S.C. Schmidt, J.N. Johnson, and L.W. Davison (North-Holland, Amsterdam, 1990), pp. 105–108.

90S01   W.C. Sweat, P.L. Stanton and O.B. Crump, Jr., SPIE **1346** (in press).

90T02   N.N. Thadhani, private communication (1990).

90Z01   J.A. Zukas, in *High Velocity Impact Dynamics*, edited by J.A. Zukas (Wiley, New York, 1990), pp. 593–623.

91A01   G.H. Miller and T.J. Ahrens, Rev. Mod. Phys. **63**, 919–948 (1991) .

91B01   M.B. Boslough, Annu. Rev. Earth Planet. Sci. **19**, 101–130 (1991).

91D01   E. Dunbar, N.N. Thadhani, and R.A. Graham (unpublished).

91F01   P.S. Follansbee and G.T. Gray, III, Mater. Sci. Eng. A **138**, 23 (1991).

91H01   W.F. Hammetter, private communication (1991).

91M01   C.E. Morris, Shock Waves **1**, 213 (1991).

91S01   D.J. Steinberg, Equation of State and Strength Properties of Selected Materials, Lawrence Livermore National Laboratories Report No. UCRL MA-106439, February, 1991.

91S02   S.M. Sharma and Y.M. Gupta, Phys. Rev. B **43**, 879 (1991).

91S03   I.Y. Song, New Mexico Institute of Mining and Technology, private communication (1991).

91W01   D.C. Wallace, Structure of Shocks in Solids and Liquids, Los Alamos Scientific Laboratory Report No. LA 12020, January, 1991.

91Y01   L.H. Yu and M.A. Meyers, J. Mater. Sci. **26**, 601 (1991).

91Y02   D.A. Young, *Phase Diagrams of the Elements* (University of California Press, Berkeley, 1991).

92A01   M.U. Anderson, in *Shock Compression of Condensed Matter—1991*, edited by S.C. Schmidt, J.N. Johnson, and L.W. Davison (North-Holland, Amsterdam, 1992), pp. 805–808.

92B01   F. Bauer, R.A. Graham, M.U. Anderson, H. Lefebvre, L.M. Lee, and R.P. Reed, in *Shock Compression of Condensed Matter-1991*, edited by S.C. Schmdt, J.N. Johnson, and L.W. Davison (North-Holland, Amsterdam, 1992), pp. 887–890.

92B02   M.B. Boslough, in *Shock-Wave and High Strain-Rate Phenomena in Materials*, edited by M.A. Meyers, L. Murr, and K. Staudhammer (Marcel Dekker, New York, 1992).

92S01   E. Smith, Ktech Corporation, private communication (1992).

93G01   R.A. Graham and N.N. Thadhani, in *Applications of Shock Waves to Materials Science*, edited by A.B. Sawaoka (Springer-Verlag, Tokyo), in Press.

Subject Index

Author Index